Endorsements

"Larry Garrison was extremely helpful with setting up the exclusive interview with the two Michael Jackson jurors who now say he is guilty. Larry was very professional to work with and I look forward to dealing with him much more in the future. Larry Garrison is a true news breaker."

—RITA COSBY, *MSNBC* Host

"It has been great working with you. . . . You are a gentleman and a great journalist."

—LANA PADILLA, ex-wife of OKC Bomber, Terry Nichols

"You are a fine journalist and a true friend."

—Ellie Cook, Juror #5 in the Michael Jackson trial

"Knowing Larry, you will be extremely informed and inspired. *NewsBreaker* gives you facts through entertaining value . . . excellent read."

—RUDY RUETTIGER, "Notre Dame #45," *Rudy*

"It was great working with you and John on the North Hollywood shoot-out. You are a true 'Newsbreaker-Journalist.'"

—TERRY M. GOLDBERG, Esq., lawyer,
the "North Hollywood Shootout"

"Larry is more than a journalist and film producer, he is an activist. He represents real hope in bringing about change for the common good . . . for all of God's creatures, both great and small."

—RIC O'BARRY, Dolphin Project, www.dolphinproject.org

The NewsBreaker

The NewsBreaker

LARRY GARRISON

WITH

KENT WALKER

NELSON CURRENT

A Subsidiary of Thomas Nelson, Inc.

Published in Nashville, Tennessee, by Nelson Current, a division of a wholly owned subsidiary (Nelson Communications, Inc.) of Thomas Nelson, Inc.

Nelson Current books may be purchased in bulk for educational, business, fundraising, or sales promotional use. For information, please e-mail SpecialMarkets@ThomasNelson.com

Library of Congress Cataloging-in-Publication Data

Garrison, Larry.
The newsbreaker / Larry Garrison, with Kent Walker.
p. cm.
Includes bibliographical references and index.
ISBN 13: 978-1-59555-287-7
ISBN 10: 1-59555-287-1
1. Garrison, Larry. 2. Television journalists—United States—Biography.
I. Walker, Kent, 1962- II. Title.
PN4874.G327A3 2006
070.92—dc22
[B]

2006022084

Printed in the United States of America

05 06 07 08 09 QW 5 4 3 2 1

Contents

Acknowledgments

This book is dedicated to Shannon Buckley, Scott Brazil, Helen Garrison, Robert Garrison, and James Garrison who have been my inspiration and who I miss dearly.

I would also love to thank my children Jaime, Lindsay, and Sean, my son-in-law Michael, my grandson Dylan, Lee, Nolan, Melissa, and my sister R. Stephanie Good, who have all inspired me to try and make a difference in the world.

Special thanks to my agent Bill Gladstone, Ming Russell, Jim Baird, David Dunham, Joel Miller, Alice Sullivan, the team at Nelson Current, and Kent Walker who helped make this book possible.

I would also like to thank you, the public, for being sophisticated enough to know that you deserve the truth and will accept nothing short of that.

1

Labels

I hate traffic. Driving home up Highway 101—Hollywood behind me and my home in Westlake before me—a thirty-mile drive takes over an hour. The one good thing about Southern California road time is that it gives a person a chance to reflect.

I had just dropped off lunch for my twenty-two-year-old daughter, Lindsay, at the movie set where she was working as a key make-up artist. She had forgotten to pick it up before work and coaxed me into delivering it. I got to the set and was proud of my daughter's professionalism. I spoke with the director for a moment and watched a take. Just as I was leaving, one of the extras asked, "So, Mr. Garrison, what kind of work do you do?" I just smiled, shook his hand, and walked back to my car.

My kids have a tendency to be a little dramatic. It's a direct result of having a father who has been involved with the entertainment and news industry for the last twenty-five years. For twelve of those years, I've been a single dad. Because of that, my kids have often been exposed to my work. And like me, they're also becoming goal-driven adults with a dash of overachiever. This troubles me when I think about my middle child—how relentless the business is that she's chosen. The money and romance of the entertainment industry are hard to resist, but it's those same traits that make it cutthroat and competitive.

Ambulance chaser and *media whore* are just a couple of the less flattering descriptions used to label me and what I do. Most jobs have titles like firefighter, CPA, or whatever. One or two words or an acronym, and that's all the explanation needed. The easy out for me is to say that I'm an executive producer for film and TV. But what I do requires much more explanation than a simple title can provide. No matter what the short descriptions are, they describe only part of what I do.

Ambulance chaser? Maybe so. Part of my work requires that I be on the lookout for people who get caught up either directly or indirectly in a situation that is so far out of the ordinary that their story becomes newsworthy. Personal injury attorneys, the other so-called ambulance chasers, have been the butt of jokes for years. Some view them as scavengers whose only purpose in life is to search out and exploit the misfortunes of others. Most people go through their lives oblivious to the workings of the civil law process, until misfortune rams into their lives and they really need help from someone who knows the system. Lawyers, I guess, will always take the brunt of jokes—until they're needed. Then they become a victim's best friend. Attorneys have to be familiar with the laws to represent their clients well, to ensure the highest possible settlement, or to successfully argue in front of a jury why their clients are entitled to compensation for their pain and suffering. Lawyers take an oath to do just that—represent their clients to the best of their ability.

My clients are of a different nature. I don't practice in a court of law; I operate in the court of public opinion. But the people I represent need me in the same way a victim needs a lawyer. My clients have often been thrust into territory so far from what they're accustomed to that the process could chew them up and spit them out without someone like me to watch out for their best interests. In much the same way lawyers help their clients through the legal system, I help my clients navigate the media machine, specifically the news media. And even though I don't always chase them, I am always on the lookout for them.

I have more than eighty other people in the field on the lookout. I call them stringers, field people, and sometimes producers, as they often work with me on producing a story for the news. They could be doctors or lawyers or anything at all. One of my most significant researcher/producers is my sister, R. Stephanie Good, Esq. As a lawyer, author, and sponsor of humanitarian causes in New York, she has brought me countless stories that have ended up in your living room via TV news. My sister does it to get the spiritual side of stories heard, but other bookers often have a different agenda. Most are money-hungry or just get a rush out of finding newsworthy stories. They scour local papers, keeping their ears to the ground, knowing that whoever is the first to find that story that makes you say "Oh, my God" out loud may be in for a cut on a film or book deal.

It has been said there are two sides to every story, then there is the truth. I am on the lookout for the people who can give you the truth of a story that has captured public attention. Although I have dealt with stories regarding some of what goes on behind the scenes at the White House and with many celebrities, most of what I cover is mainstream. It is not often that I am involved in matters pertaining to national foreign policy, and I offer less attention to people who claim to have been abducted by aliens. My work lies somewhere between the tabloid headlines and the stories covered in *Time* magazine, but everything is fair game. And yes, I must admit I love those major sweep stories that I put on the covers of magazines and bring to the attention of the world.

Maybe when asked what I do, my answer should be that I provide the public with the news stories it cannot get enough of—the kind of stories people talk about at the office water cooler, at hair salons, or over casual lunches. Most of the content is not really that important in the grand scheme of things, but everyone knows the stories because they can't resist their pull. I am not owned by any of the big news organizations anymore. Rather, I supply them with the stories viewers want to know about. If the story and conditions are right, I'll develop it into a book or movie.

The American public's appetite for news has changed drastically over the years, as has the news itself. A few decades ago, fatherly figures on sterile sets provided information on the events of the day and left it to viewers to form an opinion of what was important. They delivered the facts and the public was left to draw its own conclusions. Once in a while, the newscasters gave their opinions, but they clearly stated it was a commentary. There were no twenty-four-hour news channels, and, aside from the nightly telecasts, most of the news shows were seen Sunday mornings and on the occasional bulletins that are now called breaking stories. News organizations left entertainment to the entertainers that followed the newscast. Somewhere along the way, news moved away from the news and became more of a business, and the priority changed.

Ratings are paramount, and it is the American public that

drives the ratings war. News organizations still pride themselves for being accurate, informative, and unbiased; but the ratings war has changed the face of news and the way it operates. Instead of the Walter Cronkites of yesterday, today you see younger and much-more-attractive-than-average newsmen and newswomen reading news scripts into the camera. TV news has become more of a show. News is delivered from elaborate sets, luring viewers to believe that newscasters are sitting in a living room while they sip their morning coffee; or that the newsroom really is right there in the studio, and reporters can actually be seen working diligently in the background. Many times, a ticker tape runs along the bottom of the screen to offer additional news, just in case the story being covered by the talking head isn't good enough to keep the viewer's attention. As they deliver the news, there is more voice inflection here, little comments there—whatever it takes to keep you tuned to their channel and not the competitor's.

In Cronkite's day, all that was heard in a newscast was the voice of the reporter. Now, when sad stories are reported—the discovery of the bodies of a young boy and his mother after an extensive search hoping to find them alive, for instance—you hear mournful music in the background to accentuate the tragedy. There are also the shows that try to sway an individual's thought process as they report the news. This type can be seen just about any time of day, seven days a week, fifty-two weeks a year. These shows take current news stories and milk them for all they are worth. The subject matter might be petty, like why movie stars' marriages fall apart. But it can also be much more significant, with far-reaching effects.

Ten years ago, in December 1996, the nation first heard the

name of six-year-old Jon Benet Ramsey, who was found murdered in her upper-middle-class home in Colorado. The case soon became the subject of household conversation, due to the amount of coverage it received. Before this, the job of the news agencies was to report only the facts, but something changed with the Ramsey case. The media took it upon themselves to play jury. Instead of simply reporting the evidence, reporters analyzed it on-air and drew conclusions as to what might have happened. No one came out and said that the child's parents killed her, but almost every update to the story made sure that the public knew that the parents were the prime suspects. When new evidence was brought to light, the media would spin it to seem more damning to Jon Benet's mother and father. If evidence was introduced that might lead people to believe the parents were not guilty, it was downplayed. With the barrage of reports and the loss of true objectivity, the media, in short, passed judgment on John and Patsy Ramsey and destroyed what remained of their tragic lives. The media reports that Patsy was jealous of her daughter and trying to live her failed modeling career through Jon Benet was sickening. It was almost as if the news agencies were trying to establish a motive to support their suspicions. Reporting quickly changed to speculation and became cruel.

I spoke with the Ramseys in the early part of 1997. We discussed the possibility of doing a book to strengthen their case in the public eye, but at that time the media was on a witch hunt. Even though I brought plenty of information to the networks that contradicted what they were reporting, it was too late. The only thing they were interested in broadcasting was absolute proof that the parents didn't kill their daughter, or things that would increase the suspicion that they did. It is a lot harder to prove you didn't do something than to prove that you did. In my gut, I knew the Ramseys had nothing to do with their beautiful daughter's death. I remember the frustration I felt for them as we parted ways later that year.

In January 2005, DNA evidence proved that the Ramseys were not guilty. Scientific evidence showed that someone else had been in the basement of the Ramseys' home. The news magazine show *48 Hours* ran the report, and a few pieces did make it to the air, but the damage was already done. The only people who truly hold the media accountable for what they report are the media, and it doesn't add to the ratings when you admit that you were wrong. For every minute of coverage the exoneration received, there were hours that implied the Ramseys' guilt. Regretfully, Patsy Ramsey has recently passed away.

Maybe it was with the advent of cable news. Or maybe someone just came up with the idea that if they could make the news more interesting and entertaining, more people would watch, the ratings would go up, and more money could be made. Whatever the cause, the newscast as we knew it changed.

Even the stories, or types of story, have changed. Things that made page three of the local newspapers in the fifties and sixties are now part of the headlines. What middle-aged man from Middle America killed his wife and almost got away with it, or what teacher is having an affair with her underage student? Things that were not considered newsworthy back then, or were just swept under the rug, have become part of the headline hype. The stories that really have no impact on the day-to-day lives of the majority of viewers are the very stories that they become fascinated with and have come to demand from their news.

These stories have to have someone to tell them, and that's where I come in. I represent the people who have the news that the American public is hungry for. That hunger has created my job.

This is where it becomes a little complicated. People want to be able to take for granted that the information they receive from their news is accurate. They don't want to feel that they have to second guess the information being provided to them. At the same time, they want to get as much information as possible—at least as much information as they find interesting. The public is becoming more knowledgeable of how the media machine works and expects more details with their news. The networks know that if a story can grab the public's attention and they can present the story in a way that is entertaining as well as informative, the ratings will improve. With this in mind, they have to dig deeper and be ready to report on different aspects of a story in a way that will hold the public's interest. It is this relentless need to maintain a top position in the ratings war that creates an inherent danger of reporting something other than the truth.

Part of my job is like that of an investigative reporter/producer/journalist—to dig through the facts and make absolutely sure the information I relay to the news agencies is accurate. Literally hundreds of stories are run past me every month, and, at the risk of sounding cliché, many times I have little more than my gut instinct to filter out the truth from the myths—at least at first. I don't rely strictly on gut feelings to verify the information brought to me, but I have to admit, in the last twenty-five years of bringing the "Oh, my God" stories to the news agencies, my gut feelings are exactly what have kept me out of hot water. Not one of the pieces I have brought to the news has proven to be false. Often, it was nothing more than a gut feeling that stopped me from pursuing a story, which was later found to be a lie.

The bigger the story, the more important it is to be diligent. In April 1995, the bombing of the Murrah Federal Building in Oklahoma City, where more than 160 people lost their lives, shocked the nation. At first, the news agencies led the American public to assume it was a terrorist attack from outside the country; but when evidence revealed it was an American that was suspect, the media had to change gears, and the hunt for any information on Terry Nichols and Timothy McVeigh was on.

A few weeks after the reports first implicated Nichols and McVeigh in the bombing, Lana Padilla, Terry Nichols' ex-wife, contacted me saying she wanted to reveal information about her ex-husband. With the type of work I do and as important as this story was, I could not help but think this would be another huge feather in my professional headdress. After numerous telephone conversations, I realized that Lana and Terry's son Josh were just as amazed about Terry's role in the bombing as everyone else, and they really didn't have anything to offer to the story at that time. Because of the size of the story and Lana being who she was, I had no doubt that a book could be written from her point of view, but I passed. Ten years later, Terry Nichols revealed information to Lana, and timing became a factor in developing her second book with me.

What I did find intriguing was the call I received from

someone who identified himself as "John Doe Number Two." The man told me that he had been falsely implicated in the bombing disaster. After a little investigation, I found that his only crime, which wasn't a crime at all, was wearing a shirt that resulted in false implication. His name was Todd Bunting. After a little work on his situation, we were able to help him prove that he was innocent of anything with regard to the bombing, and he was able to go on quietly with his life.

There are different types of reactions from people who are thrust into the public eye or think they are in a position to be thrust into it. Some are just plain scared. They do not want any attention and just want the situation to pass. Some crave the limelight; they get a high out of the attention given for being involved in something people think is important. There are the ones who really do want the truth to be told; who, in many cases, will put themselves at risk to make sure it is. Others use the attention as a cleansing process, getting things off their chest that they have held inside for a long time or nobody believed. There are as many different reasons people get involved in the media machine as there are stories. The ones I watch out for are those I call opportunity seekers. These are the people who see getting my attention as their shot at making it in the entertainment business, and they are willing to bend the truth or just plain lie to get in front of the camera or see their names in print. When someone from the media wants to talk to you, you become more important. Sometimes, what the opportunity seekers say cannot be trusted. I have no problem with people pursuing their dreams or making some money. I do have a problem with people being dishonest to get there.

CBS and Dan Rather found out just how disastrous broadcasting something wrong can be. In the latter part of 2004, the negative side of the obsession to be first to report a story reared its ugly head. Getting the scoop or beating out the competition to get a story on the air is a boost to ratings. The bigger the story, the bigger the reward; but one mistake and one's credibility is lost and the news organizations become the news themselves. In the midst of the 2004 presidential election, CBS jumped on the bandwagon to discredit our commander in chief. Michael Moore's Bush-bashing movie *Fahrenheit 911* was a blockbuster at the theaters, and books were being published that cast a disfavoring light on the Vietnam service of Democratic presidential nominee John Kerry. It seemed that the ratings would improve if some dirt could be pulled up on either of the candidates' military records.

60 Minutes and its spin-offs are considered the mothers of "reliable" reporting. Therefore, when Rather reported on September 8, 2004, that he had documents and interviews providing incontrovertible evidence that George W. Bush had received special treatment and did not complete his service in the Texas Air National Guard because of family political connections, people believed it to be true. The next day, all hell broke loose.

Other media giants and several watchdog groups attacked CBS and Rather, refuting the credibility of the interviews and the documents the report was taken from. But Rather stood his ground and was adamant at times, arrogantly dismissing all accusations. Some call CBS's stand a cover-up, others just poor judgment. Whatever you want to call it, it lasted about a week and ended with a public apology from Rather. CBS had to come clean and admit that the documents used in the report to discredit the president's military career were most likely false. The media had turned on one of its own, and CBS appointed an independent panel to figure out just what went wrong.

It was a good show on CBS's part. They brought in some big hitters like Richard Thornburgh, former attorney general for presidents Reagan and Senior Bush. After months of speculation from everyone in the media, CBS announced, in a press release of all things, that they fired four top executives to whom they assigned the blame. Rather announced he would be stepping down from the anchor desk. CBS turned out a little tarnished, the news media overall is a little less trusted, and the ratings war continues full steam ahead.

One of the four executives forced to leave CBS was a top news journalist and someone for whom I have a great deal of respect. Betsy West was responsible for bringing me into ABC years ago. She is one of the most professional people I know and has always been well respected in the business. I learned a great deal from her on how to present stories that were fair and accurate. I am certain that Betsy was a victim of circumstances and had to take the fall with the others. Unfortunately, that is the nature of the business—a career of great journalism and integrity washed away by the overzealous drive of others to get a story on the air.

Ratings don't improve if you're the second to report something significant. You cannot brag in commercials that you were *almost first* to bring your viewers an important news development. CBS and Rather should have been more careful, but the nature of the business brings with it the inherent risk that something that is not really news, but a myth, will be reported. The cruel part for Rather is that his brilliant career will be overshadowed by this mistake. To add insult to injury, Walter Cronkite was quoted weeks before Rather's departure as saying that "Rather was not my choice for taking over the evening news at CBS." Cronkite viewed Rather as more of a newsman, whose main priority is to get a story on air, than a third-party reporter, who is objective. Cronkite also said later that he couldn't figure out how Rather kept his job as long as he did. CBS lost credibility, but, right or wrong, they also got some publicity, which could conceivably help the ratings in the long run.

Spinning stories is another risky way the networks battle over the ratings. Some think to put a spin on a story is to make it something it is not. It's not that simple. No one knows who was the first to use the term, or when it became part of the media vocabulary, but it has become a very real part of the business. Spin makes a story seem more important than it is, or allows a reporter to speculate to the point where nonfiction comes dangerously close to fiction—anything to add more life to the news piece that will keep viewers tuned in and wanting more information. I'm sure the term comes from the toy that was popular before video games and remote-control cars took over our children's imagination. A pull of a string and the top took on a life of its own. The energy would send the plaything across a surface, and it would continue to capture the imagination until all of the energy was used up.

Today's media has to use spin to stay competitive in the ratings game, and it has learned how pull the strings. Just like the toy, the harder the spin, the more unpredictable the story becomes. I look for stories that can be spun to stay competitive, but I must also be certain that all the information can be proven true, or at least not proven untrue. To complicate matters further, I also have to decide between what is newsworthy and the public's right to know, and what is better left out.

This is where I have to go out on a limb. As important as it is for the networks to report the news, for me it is just as important that the people who come to me with their stories be protected from the media and the spin they put on many of the stories. Some are of the opinion that the public has the right to know everything about anyone who is in the public eye. Others feel that there are boundaries as to what should be broadcast. I operate somewhere between those two positions, and they are always subject to change. It is not the stuff with which popularity contests are won.

In February 1997, I procured the story of Dick Morris, an advisor to Bill Clinton for his presidential campaigns. The events that transpired over the next few months blurred the line for me as newsman and human being. Dick found himself right smack in the middle of a controversy that was tearing his family apart. Bill Clinton's lack of self-control when it came to women was fast becoming the easiest story for the media to spin into big ratings. Even before anyone heard of Monica Lewinsky, Clinton's extramarital affairs were a mainstay of the nightly news. Then Dick Morris's private life was exposed, and his family had to endure the media's microscope as they reported on his extramarital indiscretions. Some would say that the public had every right to know about the private lives of the Morris family, given his relationship with the commander in chief. Others would say it is really no one else's business outside of the family. In any event, I found myself making the decision for the public as to what was really important for them to know. I chose to help a family rebuild.

I did everything I could by calling in some favors to protect the Morris family and keep the problems out of the media. I was not entirely successful. They were trying to save their marriage, and it was obvious that the added pressure of the media scrutiny would make that impossible. I had witnessed firsthand what media attention can do to people long after reporters have stopped chasing them and the story itself became a distant memory, and I take the responsibility of protecting lives from needless harm seriously. The networks will always feel it is important to put the spin on their material. I will always feel it is just as important that the people who come to me with their information be protected from the media, which have a tendency to jump to conclusions. I am sure there will be many who disagree, even dislike me for taking this position; but I gave up on the popularity contest long ago.

Dick Morris received counseling and his marriage was rebuilt. He published *Behind the Oval Office* in 1998, chronicling his experiences at the White House with very little spin—and without attacking his associates. The book was a success. I worked with Dick and his lovely wife. We became friends when he came to Los Angeles. I have Dick's autographed book cover and statement to my son encased with a shirt from Clinton and a pen from the White house in a shadow box in my war room. Dick wrote, "When you turn 50, I will run your campaign for President." The inscription was kind of a kick for my son, but also very inspirational.

There are no college degrees for what I do, and the only way to learn the skills needed to be successful is to do it. I didn't grow up wanting to be the guy who brings the sensational stories to the news. I wanted to be an actor. After eight years of being a successful stockbroker on Wall Street, I decided to pursue my dream, so I packed up and moved to California.

I had a good start. In New York, I studied with Lee Strasberg, who is known as the best acting coach in the world. He taught method acting, and it helped me procure a recurring role on the soap *Santa Barbara.* I also played opposite Nick Nolte in the movie *Mulholland Falls.* As my success grew, so did my understanding and love of the industry, but things changed suddenly. In 1994 after seventeen years of marriage, I went through a heart-breaking divorce and found myself a single dad with three young children to take care of. The pressures of going to auditions and casting calls and still making sure the kids were to school on time forced me to rethink my strategy. So, I looked into the producing side of films and TV shows and forgot about being a star. I quickly found that I enjoyed working with the true stories, and I found inspiration in them. The public feels the same way, which made them a much easier sell. The producing was a natural; I could work from home and still fulfill my responsibilities as a father.

In the early 1980s, I got the rights to do a story on March of Dimes poster girl Tracy Taylor. This young lady was incredible. Despite her disability, she was an accomplished snow skier, gymnast, and horseback rider. I entitled her story "A Child of Joy." It caught the attention of *People* magazine and the publication did a story on her. One thing led to another, and before I knew it, I was invited to partner with the Dick Clark Film Group. I was in awe. Growing up, I had combed my hair like Dick Clark of *American Bandstand,* and now my office was on the same floor as his. I was so hell-bent on making a good impression that I started moving my things into my office very early in the morning so I could get a fast start. I was a bundle of nerves as I rounded the corner toward my office. It looked like a janitor was on his hands and knees cleaning up a spill, and he was blocking the path to my office. I nervously asked the guy, "Do you mind getting out of the way? I have to move in so I can get to work."

A familiar face looked up at me and said, "Sure, no problem, kid." I wanted to die. It was my boss, Dick Clark. He had also come in the office early that day and had brought his Weimaraner, who had marked his territory at the door to my office. I had a flashback at that moment of being in the East School Elementary play *Around The World in 80 Days* in Long Beach, New York. I played Monsieur Le Bleu, and I said my two lines as classmates pulled a giant balloon across the stage. The audience stood applauding, and a friend of mine came onto the stage and said something I would never forget: "Larry, someday you will be a star." I think he had it backward, because that classmate was Billy Crystal. As I looked at Dick, I felt the same elation of pride and achievement and wondered if perhaps Billy had some insight for the both of us.

I learned more on the job with Dick Clark's company in a week than I had in months on my own. In my first week, I got my first mention on the front page of *Variety,* a trade paper that reports on the industry. "Garrison Moves From Quoting Stocks to Producing TV/Film" was the title of the article that neglected to mention either my success as an actor or all the years it took to get to that point.

Sometimes good situations were hard to come by, no matter

how hard I worked, but once in a while they just fell in my lap. I owned the rights to part of the story on the Lindbergh kidnapping. Richard Hauptmann's ninety-four-year-old widow, Anna, was willing to tell how Richard was innocent and reveal the facts on "The Crime of the Century." One day, some producer named Bill Self called me and said he was very interested in partnering with me on it, an offer I immediately declined. I remember asking him, "Why in the world would I want to partner with you?" He laughed and told me it might be a good idea to ask my boss about it.

Later that morning, I was in Dick Clark's office going over some of the projects we had in development and I mentioned the call. "My God, that is the grandfather of the industry!" Dick said wide-eyed. "That guy was president of CBS Theatrical." After some back-pedaling with Dick and a humbling phone call to Bill, I had a new partner for the film, and my career took a sudden turn for the better.

I can't tell the exact moment that my attention switched from the development of entertainment programs to the news side of the business, but the pairing of the two is a natural. By being a producer, I can give my clients something the networks could never dream of offering, a big paycheck. I also realized that I could be more effective as an independent producer, so I left the Dick Clark Film Group after a year and went independent. It was easier to procure the rights to stories and hype them in the news, so networks and studios would come to me, rather than the other way around. Later in my career, I affiliated myself with MTM, Mary Tyler Moore's company, as an executive producer with offices, but again I realized that being independent is the only way to go.

Some news shows will pay a few grand, or much more, for a picture or video that will entice someone to go on-screen and spill their guts for the nation to see. But if the story is credible and big enough, I can beat them out and offer the possibility of changing that same person's standard of living. Of course, not all of the stories that come to me have the potential of becoming a movie or a book. Most of the stories that warrant such exposure are often taken away by my worst enemy, public domain—a producer doing the story without owning the rights.

The only way I have to protect my clients is by controlling their news rights and maintaining independence from the news agencies. This way, I control the spin on a story and how it is released to the public. It is a game of telephone. The people who represent the news shows know exactly what I am doing and would love to keep me out of the picture; but when I secure the news rights, they have no choice but to work with me. I now enjoy a reputation for being good at what I do. Ironically, the competition often calls me on slow news days and asks, "Hey, Larry, do you have any good stuff we can use right now?"

At this stage of my life, it's not the money that is important, it's the person. I have made and lost money in this crazy business, but it is the people and their incredible stories that stick with me. Most of my clients are normal folks who get caught up in extraordinary circumstances, and in many cases find themselves at risk simply for being at the wrong place at the wrong time. They didn't want to have their lives become part of the media circus, but they do want to avoid the limelight. I have learned that if I put them first, the success will follow. By protecting my clients from the media monster, I can preserve their lives and still bring stories to the media that the public wants to see. Although money is important, it can't be the priority. This belief makes me better at what I do. Many of the people who come to me with their stories also have a spiritual investment in the situations they find themselves in, and I always look for those. It always pays off in ways money never could, and I respect the people who are not afraid to take a stand for the right reasons. I take care of my clients, and a trust is built that allows the truth to come out—then the real story can be told to you.

That is what my job is really all about, making sure that the real story can be told. I find the people with the stories, lock up their news rights, get some media play, help them make a few bucks, if possible get a movie or book deal for them, go on to the next story. On the surface it sounds simple enough, but it's really not that easy.

After the endless, traffic-laden drive, I pull my car up to the guarded gate of the community I live in. A new guard pokes his head out of the guardhouse. He smiles as he sees my license plate, "MOVIE TV." I'm never really sure what people think when they first see it, but it is fun nevertheless. He presses the button that opens the twenty-foot steel gates that act as a barrier between community residents and the rest of the world. Off to the left is a lake, and the glare is so bright off the water from the afternoon sun that I have to squint even though I'm wearing sunglasses. The homes in my neighborhood are stunning; oversized pillars emphasize the power of the residents who live behind them. Most of the homes are modern, but it's apparent that big bucks have been spent to give them the feel of old money. They are all custom, no two are alike, but they do share the air of affluence. A quick turn into my driveway and there is my house, the little David that sits among the Goliaths. My little cabin built in 1950 is hardly worthy of being guest quarters for some of the homes that surround it, but I love it. I am here right in the middle of what I cover with the news stories, but it is still my escape.

Just over the trees in my backyard, you can almost see the home I sold a couple of years ago. I never look over there. I stop and make funny noises at my pet squirrels as they run around their open ten-by-ten cage in my backyard. On the other side of my house is the chicken coop. Every morning I go and collect the eggs and feed the chickens. I am fairly certain I have the only chicken coop in my neighborhood. On the inside, my home is modest, but it does have some outstanding features. Throughout the two-thousand-square-foot, three-bedroom-one bath cabin you can see expensive paintings on the walls, Tiffany and Lalique sitting next to old-fashioned bearskin rugs that were a gift, and a dog pen.

My cabin does not give the impression that a producer lives here. In fact, the only feeling a guest would get is that the owner has not yet made up his mind on what theme to use in decorating his abode. Throughout the house are pleasant memories of my family. There are pictures of my son, Sean, my eighteen-year-old, who is an avid photographer. My daughter Jaime is my oldest child. She is a massage therapist who works on celebrities and the public. She has strong sense of what is really real in the world. Four years ago she made me a grandfather, a feeling that is indescribable. Lindsay's picture is there, too. My younger daughter is a mix between a successful makeup artist, a genius with FX, and the next Max Factor makeup creator. She is simply a kick.

One of the small bedrooms could be called the war room. The walls are filled with reminders of my career in breaking the news and fueling the ratings war. There are pictures of senators, presidents, famous people, and some shots of people that would be recognizable only to people in the industry. In every case, a picture has a story worth a lot more than a thousand words. These mementoes are a tribute to the accomplishments of a survivor who learned to handle the pressures of this business and still come out in one piece, for the most part.

Every one of these reminds me how crazy the world of the news really is, from gifts sent by presidents—in this case a polo shirt from the oval office to thank me for showing my discretion by not reporting the antics that only a president can get away with—to letters thanking me for helping the writers through a difficult time. Off in the corner there is an odd-looking porcelain figure, the Edgar Award for Best Fact Crime of 2002 for *Son of a Grifter.* The ugly little statue could easily be mistaken for a fifth-grade art class project. It serves as a reminder of an incredible story, which was made into a movie starring Mary Tyler Moore and established a valued friendship with Kent Walker. Kent was not only one of the recipients of this award, he lived the story and survived.

Right above the odd-looking figure is a window box containing some military dog tags under a cover of *Time* magazine featuring Michael Durant. His story of survival as a prisoner in Somalia and how the political process really works still inspires me. His story would later be told to some degree in the movie *Black Hawk Down.* I will save this one for later, though.

On my desk are some postcards and many diaries that

belonged to Bonny Lee Bakley. Her best friend is one of my clients and has entrusted the care of these belongings to me. Within the pages of the personal notes, a far different story is told from the grave of Robert Blake's murdered wife than you see on the channels that covered his murder and civil trials.

The letter from Linda Tripp and the pictures of Paula Jones's wannabe actor ex-husband, who would spill it all for a possible acting role, adorn a small spot.

An original photo of Jack Ruby, given to me with a theory to solve the assassination of JFK by Melvin Belli, "The King of Torts," is a one a of kind.

Every picture and object in the room has a story to tell, and none of the stories would be what is expected, but then that is how it works in my world. There is even a blank space that has a story, above the *Time* magazine cover with Michael Durant. I had cleared that little area for a recent project I thought was going to be the crowning achievement of my career, but it turned out to be just like the space on my wall. Empty.

I look at the cover of our *New York Times* bestselling book, *Aruba: The Tragic Untold Story of Natalee Holloway and corruption in Paradise*, that I coauthored with Natalee's father, Dave Holloway, and my sister, R. Stephanie Good. I reflect on bringing Joran van der Sloot and his family to New York for an ABC *Primetime* interview, and how I have been working to keep the story alive for Natalee to be found.

As I sit back in my old chair in the middle of the war room, I am thankful that the phones that are usually ringing off the hook with people on the other end of the line telling me they have the next "Oh, my God" story of the decade are quiet. It does take a lot out of a person, but when people start to understand the hows, they start to understand the whys. Here, in this room, I am surrounded by the evidence that I fought for the truth and in many cases succeeded. Then there are the ones where things didn't work out the way I thought they would. In any event, there is always the story behind the story, the spin if you will.

I sit back in my chair and think about my work. There have been so many news pieces over the years, it is hard to know where to begin. Another one of those job descriptions/labels comes to mind. I am also called a storyteller, and I do have many stories to tell.

2

The Get

I am a self-promoter and I am good at it. My competitors don't like me because they usually lose out to me on the big stories. As arrogant as it sounds, it is this attitude that has fueled my success in this business, the business of news. For every story out there, there are hundreds of people trying to find the parts to the stories that will make the story bigger, get more people to watch, and drive the ratings up. To be competitive, potential clients must sense confidence on my part. If I came across in any other way, I would be a casualty in this business before I even got started. No one wants to do business with someone who is not absolutely sure he can get the job done. It is essential that individuals feel that I am the one who can protect them from the media machine. My clients are assured that their story will be told in a way they will be comfortable with. By picking me to represent them they have the best shot at a book or film deal. I have a proven track record that I can produce for further assurance that adds to my image, and it works.

Constant change and a ferocious pace are staples of the news business. To be a successful player I must pay attention and be ready to react at a moment's notice. I am wired for this business. If there is a chance for a story to be developed, I can't stop until it is done. From the moment I wake up in the morning to the last seconds before I fall asleep at night, the gears in my head are always turning. The way I am made up helps me to think faster and clearer, and I never get tired of working. My clients are glad to have that energy on their side. Toning down the energy level when working with the news agencies is sometimes necessary, but they do say that the excitement I bring with the stories I procure is contagious.

The news that floods into your living room has many different sources. Many are from the correspondents who are on the payrolls of the networks. These journalists go out and find the interviews and take the pictures that bring the information to the public. On courthouse steps or at the scene of a major catastrophe, there are hundreds of news personnel running around with press passes dangling from around their necks or clipped to their shirts, hunting for the one stone that has not been overturned. It is this kind of in-your-face reporting that a majority of the pieces you see on the news broadcasts come from. Every major news organization has correspondents in areas that are consistently newsworthy. From the White House all the way down to the local police station, there are reporters present to make sure that any new developments will be reported swiftly.

News leaks are oftentimes used to manipulate the news to a desired end. In the political arena these "leaks" have become a valuable tool. By giving out just enough information to get the attention of the news media on a particular subject they can create an information frenzy. Our Founding Fathers could not have realized how much power they were giving to the press when the First Amendment was ratified on September 29, 1789. Somehow, someone from the press figured out how to use the power of the First Amendment to prevent him or her from having to divulge his or her sources, for the most part. Thus, "according to sources" became a mainstay and very familiar term in the news media, and it has made it pretty easy to leak information to the press.

Every day, I scan the newswire services or read through the smaller newspapers for the slightest mention of a story that I believe has the potential to become the next sensational piece the media will want to exploit. The telephone and computer are my main tools, and I must be in constant pursuit. Then there are the times when my people in the field or stringers will call me with something that shows potential. It is impossible to predict what is going to grab the attention of America. Many times it comes from the crime files, but every aspect of day-to-day life has the ability. When I get lucky and catch something that hasn't made it to the big news agencies yet, my job is a little easier. When I go after something that has already made major or minor headlines, I have to think about the competition and act a little faster. In either case, I love the chase—finding the people, contacting them, and convincing them that I am the one they should entrust their information and rights to. I call it "The Get."

Before I get the people I have to find them, and that can be a tough job in itself. Sometimes stories fall into my lap, like the Robert Blake case, but usually I have to go out and find them.

Fellow newsman Andy Rooney let out one of the secrets of the industry while testifying in a civil law suit. When asked by the questioning attorney how he found people, he replied in his famous deadpan manner, "Well, I use one of the tricks I learned early on as a young reporter. I look them up in the White

Pages." Sometimes it really is that easy. Those are the ones that bring me only a phone call away from the target. For the most part, though, I have to be more creative.

Most of the stories I find leave few clues on how to find the people involved. If I know what town they live in, I will search school records, old yearbooks, and contact friends, or find out what school their children go to and what time they pick them up. If I know where they work, I talk to co-workers. Any information is helpful in finding them, and oftentimes I gain insight into the target before I contact them for the first time. I find out what their attitude is toward the media, if they are interested in talking, or if they just want to be left alone.

There really isn't anything I won't do to find newsworthy people. If I find out they like sushi, I stroll through all the local sushi restaurants to see if anyone there knows the people I am trying to find. On many occasions, I contact family members, who for the most part are very leery of me; but when I tell them that I am interested in pursuing a book or movie deal for their loved one, they loosen up. When I tell the sibling or parent of the target that I would greatly appreciate them helping me contact the newsworthy person, enough so that they would be included in the potential windfall of the next movie of the week, I get their attention.

If I can find the zip code of the area where the contact lives, I might go to an online dating site. It is amazing how many people use these sites, and I will often come across someone who knows who I am looking for or who is willing to help me find them. I have had people I found on an online dating service drop off letters at the homes of my targets explaining why they should contact me. Since many times the stories I find deal with physical trauma, I will call local hospitals and get the information I need from an enthusiastic intern. I have also used private investigators in the past, but the majority of the time I find future clients through my own hard work. Being in the business as long as I have, I can use the media itself. I might trade exclusive story rights I have for information leading to a new contact. If law enforcement were as effective at finding people as news organizations are, there would be fewer criminals on the streets. Cops have to work around civil rights, huge caseloads, and due process. The press works under the cover of the First Amendment, and no one can hide from them. If they want to find you, they will. They have the resources and the motivation of the ratings, and the money involved makes them relentless.

There are literally hundreds of ways I can find the people from whom I want to procure news rights. It is just a matter of being a little creative and not giving up until the mission is accomplished. On many occasions, the news agencies will contact me and ask me to obtain the news rights from someone they have found. The people they are trying to interview may feel uncomfortable dealing with the media on their own, so they call me in to put them at ease. Once I do find a contact, I have to switch hats again. Instead of a private investigator, I become confidant/producer/protector.

No one will pay for something that is given for free, and the only real bargaining chip the subjects may have is the information they keep tucked away. I have to be careful with each new call. The subject may have already been contacted by someone else in the news industry and spilled their guts, and then it's usually too late for me to do anything for them. They may have had their fifteen minutes of fame, but that is all they will have to show for the unique position of having information the public wants. The most frustrating times for me are when people think that getting on one of the news shows and telling their story strengthens their position with the media and the hope of a book or movie deal. The opposite is true. Once it is out there on the news shows, it is public domain.

Anyone can take anything that has been reported and create a book or movie from public domain. I use the word create deliberately, because there are no checks in the system to prevent writers from creating just about anything they like to fill in the parts of the puzzle that normal reporting channels miss. The term *based on a true story* is misunderstood. Anyone can take that germ of truth and twist it to fit a story. Frequently, newsworthy individuals venture into the media monster on their own, certain that they are improving their chances for the entire truth to be told in a book or a film, although this is rarely the outcome.

Public domain can be one of my biggest challenges, it is also one of my best tools when it comes to "The Get." I explain to the individuals I contact how the system works and that granting all the interviews that are requested is not going to produce anything for them. The book publisher that has an author who can turn out a book in a few months will capitalize on the story of the moment.

When clients sign with me they will have a much better chance of having their story told accurately because I can protect them from the way the media loves to spin the news. I explain the financial potential for them, but also let them know that the story could be told through the eyes of an attorney or a distant relative, thus cutting them out of the loop. They quickly see how they need me to protect their best interests. It also becomes clear that I represent a better chance for a book or movie deal, for which they will be compensated.

Education is key. I never push for people to sign with me right away, although oftentimes they do. I try to inform potential clients about the process and what to look out for. They sometimes need time. I'm sure they want to shop around and see what else is out there, if anyone else is willing to give them a better deal or make more promises. I have been told by people who call me back to do business that the information I give them is right on the money and that they feel most comfortable with me. I am not entirely sure why people call me back, but I do make a point to be fair and honest with everyone I contact, and maybe they can feel that.

Even when there is not a potential for a book or movie on a story, I can still procure the news rights to a person's story and act as a newsagent and journalist for him or her. There is still money to be made, and often in the competitive business of news, the figures can be very high. I also take pride in protecting my clients. When people are initially thrust into the public eye, the media monster can be intimidating. Few things in normal day-to-day life can prepare anyone for the stress of being in the attention of the press. They can show up on your doorstep, at your place of work, or call you in the middle of the night. Reporters follow people with potential stories and wait for the perfect moment to approach them and start asking questions. The problem is that the perfect moment for the reporter may not be the best time for my clients. Some reporters like to use the element of surprise to catch people off guard, so dinner at a restaurant or walking out of the gym, anytime you would least expect to be approached by a guy shoving a microphone into your face is often the time of choice for reporters. It makes better news, people are more agitated—in many cases just plain scared—which makes the interview more interesting for the public to watch and easier for the networks to spin.

Many investigative reporters take pride in the fact that they only care that the news story be reported and that they will stop at nothing to make sure it is. Unfortunately, many times, it does prove to be a frightening experience for my clients. Not all stories motivate the news media to invade people's privacy, but the ones that land in the tabloid genre tend to be the ones on which the reporters take off the gloves. In the Adam Sandler movie, *Mr. Deeds,* actor Jarred Harris plays the role of Mac McGrath, the ruthless news producer who will stop at nothing to get the attention-grabbing story. In the movie, McGrath sends his reporter, armed with a hidden recorder and an accomplice in the wings to capture the video, to create situations with Sandler's character. McGrath then spins the truth and manipulates the videos to present a story that is completely false, but sensational enough to increase ratings.

The truth poses little concern. Weaving an interesting story to captivate the viewer is paramount. In one segment, Mr. Deeds goes into a burning building and saves a bunch of cats from the inferno. He then trips and falls on top of a large woman, who was grateful for his heroic act. McGrath, through the magic of video editing, broadcast a story that made it look like Deeds killed a bunch of cats by throwing them out of a third story window, then went on to molest the woman. Although the story line in the movie is a little exaggerated, it does show the mentality of some of the news shows out there. Creative editing is used to distort the truth, and protection from such situations is one of the essential services I provide for my clients.

Many people think it is in bad taste for individuals like Amber Frey, Scott Peterson's infamous mistress, to capitalize on the information they have. Even though she claimed that Peterson lied to her from the start, many believed that the murders were motivated by their relationship. Meanwhile, Lacy Peterson and her baby are still dead, and the family is overwhelmed with grief. Some were incensed Ms. Frey had the audacity to make money from such a horrible situation. Public opinion aside, the fact of the matter is that it sells books and movie tickets.

What is outrageous is that the people who actually commit and are convicted of crimes can realize a profit from selling their story to media outlets. In the 1970s, New Yorkers were terrorized by what was then called the ".44-Caliber Killer," later known as the "Son of Sam." Son of Sam, whose real name is David Berkowitz, was offered big money for his story rights after he was arrested. People were shocked. It didn't sit well for many that convicted murderers could make a profit from their crimes through the media. In response, many states passed laws that made it illegal for anyone convicted of a crime to make any profits stemming from those crimes. The laws are known as Son of Sam laws.

New York was the first state to pass a Son of Sam law, and it was also the first to have the law struck down by the U.S. Supreme Court in 1991. In February of 2002, California's Son of Sam law was struck down by the state Supreme Court, citing free speech concerns. The same amendment that gives the press free rein in pursuing and reporting the news also protects the free speech rights of convicted serial killers. Although Son of Sam laws have been found unconstitutional, lawyers have found ways to prevent criminals from making a profit from their crimes through the civil process. The best-known example is the $33.5 million judgment a civil court in California levied against O. J. Simpson in 1997. When all things are considered, the chances of anyone convicted of a felony making a paycheck because of it are slim.

For those who are involved, but are not guilty of wrong doing, the financial rewards can be a lifesaver. In the early '80s, John Delorean was found innocent of drug-trafficking charges. He was the golden boy of the automobile industry in the '60s and '70s and the youngest executive vice president of a U.S. auto company in history. He loved the limelight and spent as much time in Hollywood as he did in Detroit. In the '70s, he left General Motors behind and started a company to develop the sports car that carried his name. At first it looked like the company was going to be a success. The car was well received by the auto industry and the public, but after a couple of years things changed. The cars were being built in Ireland. The local government was heavily subsidizing the enterprise in hopes of making the Delorean a world-class car. However, demand for the futuristic two-door, made famous by the *Back to the Future* movies, was far from what was needed to make the company a success. Along with the Irish government, Delorean had some powerful people invest in his company. The *Tonight Show*'s Johnny Carson was one of the biggest. It was from behind the wheel of a Delorean sports car that Carson was charged with drunk driving in 1982.

When John Delorean was charged with attempted drug trafficking, it shocked the nation. The news reports made it easy to see how the one-time giant of the auto industry had planned to traffic millions of dollars worth of cocaine to save his defunct car company. There was videotaped evidence, conveniently leaked to the media, of undercover FBI agents locking up the deal in a Los Angeles hotel room. At the end of the tape, there was Delorean being read his Miranda rights. It seemed to be an open-and-shut case; motive, opportunity, and convincing evidence. Then the trial told the rest of the story.

Delorean's attorney, Howard Weissman, convinced the jury that his client was a victim of overactive undercover agents. Larry King was later quoted as saying, "It was the clearest case of entrapment I have ever seen." By the end of the trial, Delorean was completely vindicated and found innocent of all charges. That is when the news coverage stopped. What wasn't reported was that the tab for Delorean's defense was enough to wipe out the budget of a small city. Here was a guy that twelve jurors found not guilty, as did most of those familiar with the case. Still, the cost of proving that innocence was everything he owned; his home, money, and just about everything else of financial value was eaten up by legal fees. His car company was gone, and the only way to put his life back together monetarily was to hit the book circuit.

The people I contact often find themselves in positions of financial hardship for a variety of reasons. Being the focus of attention for the short-term usually doesn't do anything to put food on the table, but many times the situation can prove to be economically and emotionally devastating. Lives are turned upside down.

When I first contact a prospective client, I let them know who I am and what I have accomplished in the industry. I explain to them my history as a consultant/producer for ABC news and Time Warner news. I mention names of people I work with, like Barbara Walters and Diane Sawyer. I make sure they know that I co-executive-produced *Like Mother, Like Son,* the 2001 made-for-TV movie starring Mary Tyler Moore as the infamous Sante Kimes (more on this story in chapter seven). I explain to them how the media monster works. "No, your story is not worth a million dollars." I also have to scare them a little about public domain and the fact that media outlets really don't care about them one bit. They know they can hide behind me if they need to, or I can get them on the news shows if need be. It all depends on the situation. In some cases, they need money right away and a little exposure would be helpful for the long-range goals. In others, it's better to wait in the wings and just stay out of sight. I become part confidant and part coach.

There are also times when the people I try to procure news rights from are already represented by legal counsel. Most of the time, if I want to get to those people, I have to go through the lawyer, and that is when the "get" is no longer fun. Higher-profile cases attract a lot of attention from the media and there are plenty of lawyers who love the opportunity to strut their legal expertise for the camera. The higher profile the case, the more likely an attorney will represent their client *pro bono*—the legalese way of saying *for free*—and very often the motivation for the lawyer to represent the client is media exposure. I have contacted attorneys who are representing individuals whose stories I want to procure the news rights to and had the lawyer offer to sell out their clients for media exposure or the chance for a book or movie deal.

The late Mr. "If the glove doesn't fit, you must acquit" Johnnie Cochran became a household name with the O. J. Simpson case, and plenty of attorneys took notice. Egos are stroked and name recognition increases when the press gives lawyers attention outside the courtroom. There has always been a certain amount of limelight shed on the legal counsel in significant cases; but as television news has changed and trial coverage has become more in-depth, so has the potential for the lawyers to become the stars.

They may be representing their clients in the legal proceedings, but they are most often out for themselves in the media. In fact, I have come to realize that there are two types of lawyers out there: those who are truly focused on the well-being of their client's legal position and those who look at their clients as their ticket to fame. The former will almost never even let their clients talk to me, let alone the media, unless the added attention can help their case. The latter will make sure their clients do talk, as long as there is something in the deal that will add to their notoriety or pocketbooks. I have also learned how to use the latter one's greed against them, if need be. By having me as part of their team, the clients have the added advantage of having an expert in the media on their side, and I have found that when I am looking over their shoulders, the lawyers are more focused on doing a good job for their clients.

In 1998, I wanted to procure the rights to the story of Sante and Kenneth Kimes, the mother and son who became the tabloid headline of the decade in New York City. They were accused of murdering a wealthy widow, whose body was never found, in an attempt to steal her $8-million mansion in midtown Manhattan. As reports saturated the papers and airways, Sante and Kenny's history became a story too hard to resist. Murder, slavery, and arson just for starters. Mel Sacks was representing Sante, and it became clear to me within minutes of contacting him that his motives were media-driven. In our initial conversation, he laid out his plan to have the movie made from his perspective, and said that all financial profits from the case would be his. Another attorney on the case was Matthew Weissman, and he was a little less subtle than Mel. He started our conversation by saying, "The one who gets the news rights on these two will be the one who takes the best care of me." Matt was later convicted and sentenced for fraud. It seems he tried to collect millions from the accounts set up for the families of the World Trade Center's victims after the 9/11 terrorist attacks. The problem was, he wasn't representing any of the survivors. Last I heard, he was still serving time.

The Kimes' legal team did get their fifteen minutes of fame. There were plenty of interviews and publicity in the major newspapers. They even made it to *60 Minutes,* where Steve Kroft interviewed the mother-and-son crime team. Before *60 Minutes* ran their report, the pair had been portrayed as pathological liars, and Sante was known to assume different identities in her scams. Not exactly the easiest of clients on which to build a case for character. When the reporter mentioned that Kenneth got an "A" in his college acting class, Weissman screamed, "Cut!" and the interview was over. He didn't like the way the question was poised to relate acting with lying. The report used the spin on the lawyers as much as it did on the mother and son grifters. These are the types of lawyers I try not to work with. They are too unpredictable and will almost always ignore any agreements with regard to news rights by looking for other deals. I do, however, enjoy working with the attorneys who truly have their clients' best interest at heart.

News agencies have people on the payroll that do much of what I do. They are out there looking for the same stories or the same people with the stories. They are called bookers, and instead of finding people and procuring their rights for future ventures, they are on the hunt for exclusives for the news shows. Every news program loves the word *exclusive*. This powerful word means better ratings, more credibility, and glorified bragging rights. To be the first and only news outlet to bring a story to the public is one of the most important aspects of the business. But, once the story hits the air, everyone else in the business will be jumping on the news piece right away. So the glory of being the sole agency reporting the story, or getting the scoop, is a very fleeting and sometimes false satisfaction.

Many times, bookers will contact me to lock up the news rights on someone, with the understanding that they will get the first exclusive. Part of my job when representing a client is damage control, and I have learned the hard way not to grant an exclusive to anyone for a long period of time. It is dangerous for any one news agency to have the exclusive, as there are no checks to prevent them from forming any type of spin. Once everyone has a shot at the news piece, it becomes a little safer, as the competition will flush out any spin that is not accurate. The TV news magazine *Inside Edition* reporters tried to get me to do news pieces with them, then went behind my back and tried to cut me out of the loop with my clients. The only real check I have with shows like this is that in the ever-changing world of news, all the shows may need what I have one day, so most have learned to play respectfully.

When I first started in the business, everything was fair game. Find the story, get the rights locked up, and go for a paycheck. It didn't matter what the story was about or who it involved. I had a single-minded focus to get the job done, and I wouldn't stop until it was. I was good at it, so much so that I often got calls from media outlets requesting that I go out and lock in the "get" on certain stories.

In 2001, I got a call from a recently-hired booker on the news show *Extra*, and when I hung up the phone I felt sick to my stomach. An alligator had eaten a small child in Florida that day, and the people at *Extra* wanted me to contact the parents to get a quick, exclusive interview. I didn't have the heart to honor the request; in fact, I let them know that I couldn't believe they would even imagine trying to get the parents to spill their tears for an interview at a time like that. But that is what part of the news business has become. People don't even have time to mourn before they are paraded on the news shows to exploit a tragedy or become vocal about a cause. I did enjoy working with the people at *Extra,* and considered them a class act, but they needed fillers on the slow news days to keep up in the ratings. Supplying the fillers is part of my job, along with doing the major sweeps stories. The ratings wars know no emotion and never, ever feel any remorse. I draw a line on what I will go after, and I have no interest in exploiting anyone's misery. Sadly, there are plenty of people in the industry who do.

Sometimes though, I will work on stories that would be

considered distasteful. In my heart, I know the story really should be told and will accept the assignment. An attorney once called representing the Eldridge brothers, who had gunned down their father in Arkansas. He wanted to present the case that the boys were severely abused by their father and thus justified in killing him, and the media were starting to join in on the spin. I did procure the rights and brokered the story to the different news organizations, but that is where I stopped. I have no doubt that it would have turned into a successful book or movie, but I chose not to pursue it. I saw the value of the content of the story for the news but was uncomfortable developing a story for distribution that would condone murder. I have kids, and I am just as concerned about what they are exposed to as any other parent.

Credibility must be carefully guarded by news organizations. Once it is compromised, they are considered unreliable, and the public will look elsewhere for its news. One way to protect credibility is to have a no-pay policy, at least publicly. The thought is that if someone is paid for information, that information may appear to be tainted. If the public thinks someone has been compensated for an interview, the perception is that the person is more likely to lie. It is also the perception that if a person presents facts without compensation, that the news is truly news and not a fabrication. I have kept an unofficial record over the years, and whether or not someone has received compensation has no bearing on the chances of what is reported being true. There are money-hungry people who will say anything to get a piece of the limelight and try to cash in on a fabricated story, but they are easily weeded out. Their stories have a tendency to grow with each retelling.

The bookers at the news agencies need to have someone involved that can offer more than they can, and many times that person is me. I may have the ability to pull in a story that is out of their reach. Sometimes the employers do not even know about it. Bookers are not paid very well, and they see a bigger paycheck if they take a cut from me on bigger stories. They also need me to help them keep their jobs. If they do not produce consistently, the networks will drop them. So, they call me in to get the story they can't. The "get" for me is about the same as the ones I have to cold call. The only advantage is the added credibility provided by the recommendation of one of the news shows.

In 2002, a booker from *Extra* called me. Most of the time, when they needed me to procure a story it was fairly cut and dry, a little sidepiece for a bigger story that they felt my expertise would lock up for them—but this one was different. In fact, it turned out to be one of the biggest "gets" of my career.

A couple of weeks earlier, Robert Blake's wife was murdered outside a Los Angeles restaurant. The media learned their lesson with the O. J. Simpson case: When a celebrity is suspected of committing a felony and it is reported, people watch. With that case, many of the "news" talk shows spent the entire year covering only that case, and the ratings were high.

Blake's wife, Bonny Lee Bakley (a.k.a. Bonnie Lee Bakley, a.k.a. Leebonny Bakley), was sitting in their car alone outside and down the block from the restaurant where the couple had just had dinner. According to Blake, who was the last person to see her alive, he had returned to the restaurant to retrieve a gun he had forgotten in the booth where they had eaten. He explained that his wife had been receiving threats and he was concerned for their safety. When he returned to the car, he found his wife had been shot twice and was bleeding to death. The specifics of the crime made it irresistible for TV news. There were no witnesses to the crime, no confessions, the victim was married to a well known actor, and no evidence that anyone knew of made it an open-and-shut case.

The LAPD and Los Angeles District Attorney's office had learned their lesson in the O. J. case as well, and they were keeping tight-lipped about any specifics. They did make a point to mention that Robert Blake was not a suspect in the murder. So much so, that it was clear that Robert Blake was a suspect. It was the kind of story that spin was made for. So the media speculation machine kicked into full gear and dug in to find out as much as they could about the marriage, Robert Blake himself, and the victim. They soon found that Bonny Lee was no ordinary victim.

Reports were soon broadcast about Bonny Lee's questionable past, and people were coming out of the woodwork telling their stories of how she was nothing short of a full-fledged grifter. People who knew her when she was young described a girl focused on getting the attention of movie stars, and she would stop at nothing to get it. Other informants divulged that she sold nude pictures of herself and used alias after alias in her dealings with the famous. She claimed that Jerry Lee Lewis fathered her first child, and she also hung around people like Marlon Brando's son, Christian, who had a history of problems with the law.

When I first received the call from *Extra,* the booker asked me to lock in someone who had more information about

Bonny Lee's past. I was a little puzzled, as I really thought that all the media exposure would have already covered any of the interesting details the story had to offer. He then told me that they had someone named Christina Scheier, Bonny Lee's best friend and confidante, who had postcards, diaries, and letters chronicling thirty-five years of Bonny Lee Bakley's life and grifting.

The diaries and postcards that Christina kept over the years from Bonny Lee were like whispers from the grave that screamed to be heard. The more I talked with Christina, the more I realized how powerful her story really was, and I procured her news, film, and literary rights that day. Christina really wasn't that interested in the money; she wanted to tell what she knew to be the truth about her best friend. Very soft-spoken and a little shy, Christina did repeat some of the stories that were already out in the media and had even more to offer to tarnish the reputation of her lifelong friend. But she also brought a personal touch, in the way only a close friend could. Between the stories of how Bonny Lee bilked lonely guys out of money in online dating scams and the several marriages, Christina added a more human touch: some of the whys and hows, along with insight into how Blake reacted. This material warranted the development of a film and book. I immediately went into gear, placing Christina Scheier and myself on major shows like *Larry King Live* and numerous others.

No matter how I find the people who have the stories, or the details of the stories that will fill out the headlines, I have to be careful. I need to be objective, but the involvement I have with my clients is far different than what the news organizations have with them. They get the facts, conduct the interview, and move on to the next story. For me, a relationship is developed, as I am in contact with clients for months, sometimes years. Oftentimes a bond is formed. Not in all cases, but in many, I start to feel for the people and get a true understanding of the stories they have to offer. The news shows can be superficial in the way they cover a story and only report on the parts that will keep the public's attention and improve ratings. There is simply not enough airtime to tell the whole story. In my unique position, I have to know the entire story, not just the parts that are headline-worthy. Even when one of my projects becomes a movie or a book, many aspects of a story can be lost because of space and time constraints. Business is business, and it is part of my job to decide what is to be brought to the public and what is to be edited out, but often what is left on the editing room floor is what gives the story life. Over the years, I have seen stories ranging from survival to suffering, and knowing all the facts of what my clients have been through or seen gives me the motivation to do the best job I can for them.

When I take over for my clients, I can open a window of opportunities that the news agencies cannot. I am independent and have access to all the news media. Having someone like me in the loop provides more creative ways for the news shows to not pay for the news. Make no mistake: News is big business. Interesting news pieces mean better ratings; better ratings mean advertisers pay more. And then there is the competition. At any time during the day there are at least a dozen news shows playing simultaneously, even more during prime time. With this, and the fact that the public's appetite for news is insatiable, news organizations have had to become more creative in the ways they deliver the news and remain competitive. The ratings war rages on. Yesterday's news is old news, so the networks are endlessly searching for fresh stories, seven days a week, 365 days a year. Money is the necessary commodity to stay competitive and the networks' checkbooks are very, very big and always open.

3

Show Me the Money

I spent ten years as a successful Wall Street broker, so I know a few things about what makes a company profitable. It really boils down to basic economics. First, a company has to provide a product or service that commands a profit. Then, throw in the law of supply and demand. The more demand for the profitable product, the more successful the company becomes. If supply is low and demand is high, the profit per product can be increased. The rules are the same whether the product is toothpaste, cars, or advertising airtime. The advertising spaces that networks sell to their customers are an extremely profitable product. As far as supply, you can't manufacture more airtime. The combination of a high demand for a profitable product that is in limited supply is what makes for financial giants. The ratings are what determine how much the media giant's product is worth.

Within the networks, news shows have a unique position in that they cross demographic lines. If you are a retired CPA, the chances of you having your TV tuned to an MTV reality show are about the same as a seventeen-year-old having his tuned to Home & Garden Television. But no matter what the age, profession, or gender, television news has the potential to capture your attention. Advertisers know this and spend millions to have their commercials aired during the news broadcasts or highly rated news shows.

There have been several books and articles critical of the news media written over the last few years. In most cases, these publications will have you believe that it is political views or social issues that influence what and how the news agencies report. They talk about some of the big players in positions to make decisions in the business and make the argument that it is their social values that impact their choices. I cannot completely disagree with these points of view. But, after spending the last twenty-five years working with every major news organization in the country and seeing what really drives this business, I can tell you the one thing that dictates how decisions are made in TV news: money.

People will always watch the news. Other types of programming come and go. Yesterday's money-making sitcoms have been replaced with today's reality shows. No one knows for sure what will come along in the future to steal the attention of viewers away from reality shows, but one thing is for certain: they will watch the news. Furthermore, whoever can bring the news to the most people will make the most money. At gyms, bars, airports, just about any public place where there is a television, chances are it will be tuned to a news show. The only real competition TV news has for the public's attention are the sports shows. Studies show that the majority of Americans will have some form of news show tuned in at their homes on a daily basis. Whether in the morning while getting ready for work, in the evening while preparing the evening meal, or on the bedroom television while you are falling asleep—chances are you will watch some form of TV news every day. It has become the main source of information and our window to the world.

These airing schedules have taken several different forms. The major networks have the morning shows that spit out the headlines every half hour then go into more detail on some of the bigger stories of the day, with quick interviews from some of the players. For more entertainment value, they will fill in the time with segments on how to lose weight or the best way to decorate the holiday table, and have a variety of guests ranging from authors to experts who tell you what mountain bike to buy.

The network evening news shows and twenty-four-hour news channels focus mainly on the big stories of the day. These shows go into a little more detail, but whether or not they go into any depth depends on whether it is a slow news day or not. Their focus is to put before you the facts of the important happenings of the day. But these shows have really faded as the public craves more in-depth coverage. In an effort to recapture its slipping ratings, CNN has changed its prime-time lineup from the headlines to shows that focus on the entertainment industry and talk shows. NBC, CBS, and ABC are all finding it harder to keep viewers for their evening newscasts and hint that changes are on the way. When Dan Rather left the anchor chair at CBS, the network speculated openly about changing the format of their news broadcast. FOX news has left an impact on the news industry by providing more sensationalist news coverage. This type of reporting has helped FOX obtain better ratings, and the competition has noticed.

By being masters of spin and sensationalism, *Inside Edition* and the recently cancelled *Current Affair,* draw in their audience.

In these tabloidesque shows, anything is fair game, and it is just as competitive as with the major news networks and sometimes very ruthless.

The public's demand has created a market for news pieces to fill the airtime for all of these different news programs, and I am the one who brings the product to the news producer's market. This is why some people in the industry call me a "news broker." After I lock up the rights to a news story, the news organizations have to go through me to get the information on-air. Every major organization will swear that they never have and never will pay for news pieces, which technically may be true. They do not buy the news. But, they have figured out how to compensate without paying directly for the information.

An all-expenses-paid trip to New York, complete with limo, top-notch hotel, and dining accommodations oftentimes is extended to my clients. Once in a while, news producers may even throw in a Broadway show to help the individuals they want to interview pass the time comfortably. For many, this will be their only shot at ever experiencing this kind of life. Being treated like a VIP in a usually normal life can be a reward in and of itself. Sometimes just being the center of attention is ample.

Pictures or videos that are relevant to the news piece can prove to bring additional financial compensation for clients as the shows will pay for the right to air them. The bigger the story, the more the pictures can be worth. But, in most cases, it is not the pictures they are really after. If someone has pictures, but nothing to add to the story, the pictures are usually worthless. If the person who is in possession of the pictures provides a solid interview and adds to the story, the value escalates, even if the picture is unclear or has little to do with the content of the story. You have seen this process at work before. The TV screen will be full of the person being interviewed and then switch to family photos or videos that look like they could have been pulled from the pages of any scrapbook in America. The shows can say they did not pay for the interview, but my clients walk away with a check.

I have also seen news shows give people brand new, expensive, professional video cameras to record events related to on-going stories, with no expectation of it being returned. That happened on a conjoined twins story that *Extra* wanted footage of desperately. They didn't write a check to the person for the information, but a camera worth a couple of thousand dollars could be considered compensation by everyone except the news organizations.

The network shows do not do this nearly as often as the syndicated shows like *Extra* or *Inside Edition.* Shows like *20/20, Dateline,* and *48 Hours* have the resources of the network news and are often able to find the in-depth information without having to open their checkbooks. The syndicated shows are different. They do not have huge news staffs at their disposal, and it is simply easier to compensate in order to be competitive.

The big network shows also have a little-known advantage. They don't actually *write* the check, but loopholes are found to cover up payments. The same companies that own the networks own many of the syndicated shows. So when a tabloid news show pays for the right to air photos from a news subject, the producers of *48 Hours, 60 Minutes,* and all the other major news shows may have access to the pictures, and to the content of the interview. The check is drawn from an account that has the tabloid show's name on it. The big shows can protect their journalistic integrity, and truthfully say that they do not pay for the news items they have reported.

Network and syndicated shows will try to entice people with claims that they can help them get a book or movie deal, even if there are none to be made. The producers neglect to tell the people about public domain and will lead people to believe that if they spill their guts on their show, that publishers and studios will be knocking on their door to put a deal together. In these cases, the networks don't pay a cent but still use the lure of money to pull a story in. Once the interview is completed, all contact is over, and rarely are there calls offering a book or movie deal. One of the most important things I do for my clients is to protect them from this practice. I have learned how to use the media by feeding them relevant bits and pieces of stories here and there, never the complete package. I catch the attention of the publishers and studio executives, enticing them to do a project with me without adding to the risk of someone doing a film on public domain. This also allows me to protect my clients and their stories from spin.

Some of the news you watch is from good old-fashioned reporting. The producers are not sitting in the newsrooms with their checkbooks open, waiting for the phone to ring with the next news piece. Once in a while, a story comes along that the producers know has a good chance of improving ratings, and sometimes the people with the stories need a little financial encouragement to tell it. Then there are the "Oh, my God" stories. These are the pieces that the news producers have to get on their shows if they want to keep their jobs. If the competition gets the "Oh, my God" story of the day, and your show doesn't, your ratings will go down, as will the value of your product. The news industry is business, and many more people have been fired for a decline in profits than for reporting something that was inaccurate.

Just like the smaller stories, not all of the big ones require compensation. But when they do, the figures range from what amounts to filling a living room with furniture to paying cash for a house. And a nice house, at that.

Baby Jessica is a perfect example of a major story that was brought into the spotlight by the players in the news industry through financial coaxing. She was the beautiful little girl caught in a bitter custody battle that captured the attention of the public in the 1990s. Hers quickly became one of those "Oh, my God" stories I mentioned before, and all of the news organizations scrambled to get an exclusive. At one point after the story hit the headlines, baskets of diapers and baby toys wrapped up and put in a crib for her younger sibling showed up on the doorstep of her house. Each of the gifts had cards from the news outlets with a number to contact them. Her front yard looked like the baby section of a small department store. In the end, a tabloid and a network ended up with the exclusive in exchange for money, plus toys.

Many times the big stories become blockbuster movies, like *Black Hawk Down.* It was based on the experiences of Michael Durant, an army aviator who was shot down in Somalia. I was influential in getting Michael placed on the cover of *Time* magazine. The heightened publicity of the situation forced the White House to get him out. Before the filming of the movie had begun, he became the focus of the media monster. The story was full of human suffering and political intrigue, perfect for increasing the ratings of the lucky news show that got him to tell his story. The news organizations' checkbooks were scrambling to get him to talk, and the syndicated show *Inside Edition* was one of the highest bidders. They shelled out to get the exclusive. I know. I saw the check.

Even though I provide the media outlets with many of the stories they report, they have a tendency to disclaim my position in the news cycle. In the July 23, 1998, edition of *TV Guide,* J. Max Robins wrote an article entitled "News Breakers vs. News Brokers." He called my line of work a "controversial cottage industry." The article mentioned two of the stories I had procured the news rights to: The Angel of Death, Efren Saldivar, who had admitted to dozens of mercy killings in Los Angeles area hospitals; and a piece involving guests on the *Jerry Springer Show* who claimed they had been coached to lie. The article was correct in that these were my clients. However, the "Angel of Death" label was a result of an interview for ABC's *20/20* done in my backyard. The film crew set up for the shot, and when they were testing the video equipment the producer looked into the monitor and saw an ornamental bird feeder with angels I had as part of my patio furniture. Thus, Efren had a new alter ego.

When Robins contacted me, I was straight with him and told him to feel free to use my name, which he did. I trusted the guy. He also interviewed news executives from the major networks. He quoted those who "would only talk on condition of anonymity." They went on to say that they would never pay anyone for news in fear of risking their journalistic credibility. One unnamed exec from ABC did admit to the pictures-for-profit deal, and the $12,000 consulting fee CBS paid to tobacco industry whistle-blower Jeffery Wigand. Other than that, the article was mostly an attempt to bash my line of work and give an opportunity to the news bigwigs who would not give their names to swear that they don't pay for news. The article did have an impact on the industry; in fact, all hell broke loose. Everyone in the business was running scared, and the networks had to figure out more creative ways to not pay for the news.

Being the victim of spin in the *TV Guide* article stung. Spin or not, I have been paid for bringing stories to the news media. According to the news shows, I am a producer and a consultant but never a news broker. I have delivered top news pieces to the doorsteps of major networks and syndicated news shows, and they never once balked at paying. The business is too competitive to let the top stories go to a competitor, and it would have been bad business on their part if they had. It was also a smart move on my part, as I used the funds I received to develop the stories into films or books.

By paying me for the services I provided, calling me a producer or consultant, the news organizations could carry on the illusion that they were not paying for the news. In this way, the news agencies sounded more newslike, and it also made it appear as if their staff reporters had retrieved the information themselves, rather than bidding on the open market. The memo line on the checks I received from the news shows would often say, "Producing Fees." But, when the story was aired and the credits rolled at the end, my name wasn't mentioned. It didn't bother me that I didn't receive recognition in the credits. I have always been more focused on representing my clients well and am persistently searching for the stories that can make a difference. Having my kids see one of my stories hit the air and say, "Hey, Dad, look! There's one of your stories!" is enough credit.

Another way for the networks to pay me is to offer me first- and second-look deals. I developed a history of bringing in the top stories and the news shows took notice. To make sure they would have the best shot at getting my next big story, they paid me handsomely to bring it to them first. I got paid for a first-look deal, then for "producing" on the same story many times. I loved the "per story" bonuses I received when my career was flourishing.

The real product the news agencies have to sell is their advertising space. The stories I provide them makes that space much more valuable. If any other business had the opportunity to increase the profitability of their product by millions by making an investment they would be foolish not to do it. The news media does.

I was still involved with the stories in other ways, also. Most of the time, I would be present during the interviews of my clients to add comfort and moral support. I was there to protect them by not allowing the interviews to be spun in a way that could be destructive to them. There were even a few times I went on camera with them; to hold their hands, so to speak. From *Larry King* to smaller shows, I was there protecting their future. On many occasions, for the smaller stories, I never even left the comfort of my home. All the work was done over the phone. I contacted the newsworthy person, locked up the rights to their story, and then called the news shows to sell them what I had to offer. If the deal came together, I received a nice check in the mail within a few days.

Despite the emphatic denial from the news industry, my profession does exist. Networks make big money from their advertisers, and they have to fight in a cutthroat business to get the major stories on the air. It is the nature of the business that has created a market for my services. It is also the nature of the business and the success I enjoy that created my competition.

At least one small outfit wanted to copycat me. They thought they came up with a new idea of going out to the nation's smallest newspapers and finding stories to sell. Then they went out and shopped the stories to the news agencies. They also sold a monthly subscription service to media outlets that listed the stories available. It must have been a news producer's dream: Get the monthly menu and order up the special of the day to get on the next show. Journalistic credibility at its best. This company contacted me on several occasions to try and get me to sell some of my smaller stories. In fact, there was a time when they called me every day. I turned them down on every occasion for two reasons.

One, they weren't even close to being in my league, and I feared that my name being connected to theirs in any way might damage my professional status. Two, I would never sell out my clients. I admired the company for their persistence, even more so when they approached NBC for a first-look deal and claimed to have gotten it. But wasn't that a conflict—making films on the movies the network would possibly get for news also? Rumor on the street had it that one of the bookers on NBC's payroll was the one to set up the deal. He was some guy being paid thirty grand a year to bring the stories in. He must have thought it would make his job easier, but I believe that is one of the things that has made the mainstream news shows more tabloidesque. For the public, it is addictive.

Not all the news that comes to your TV is from people like me who have developed a way to market information to the news organizations. The networks, for the most part, have earned their journalistic credibility by hitting the pavement and using good old-fashioned reporting techniques. When I do bring a news piece to them, they use their vast resources to make sure that everything they report is reliable, so the news you see is trustworthy, for the most part. Their journalistic credibility is protected.

Once in a while, though, something so big will come along that the networks will stop at nothing to get the exclusive—for instance, when ABC landed the exclusive Monica Lewinsky interview about her affair with President Bill Clinton. The word on the street was that ABC landed the interview by paying the BBC, England's major television network, around $250,000 for the exclusive rights to air the interview. It seems the BBC had already locked up the rights for that particular interview. Right or wrong, the public still watched, and the network's rating went up.

The most important part of my job to me is to represent my clients to the best of my ability, but I do have to be a businessman also. I have to look for the stories that the public wants to hear about and that the networks are willing to shell out money for. I have an advantage in that I do not appear as intimidating as the reporters wearing the press passes. My press pass stays in my back pocket. It is easier for people to relax with me knowing that when they talk with me it is "off the record." Without people like me in the picture, the chances of getting the behind-the-scenes, "Oh, my God" stories on the air would be far less. But, it is a double-edged sword.

People who become newsworthy have become more savvy about how the system works, and when I try to procure their rights, they often have delusions of grandeur. Business is business, and even the biggest news stories are not going to make the subjects millionaires. Far from it. It is hard for the people I contact to understand this, and that fact has become one of the most frustrating parts of my job. I try to help them realize that they may be the top news story today, but tomorrow there will be another.

News shows love to twist stories. It is curious that after spending so much money and bragging about journalistic integrity, that the stories I supply them would be reported inaccurately. People love to stretch and exaggerate. I know I said that the networks make sure the information is reliable. But in the news business, reliable means you can't prove it is untrue. You can check them yourself; look at the newswires on the Internet, then look at the way the stories are reported on the news. Whether the news organizations find the stories themselves, or if they pay me for bringing it to them, if a story can be spun, they will spin it.

Whether or not it is wrong for the news agencies to pay for part of what they report is questionable. It is challenging for them as they are simultaneously trying to remain competitive and provide viewers with informative broadcasts. The networks don't control the ratings, the public does. Although there is controversy about the Nielsen Ratings system, it is what will determine what is watched for the foreseeable future. If the public didn't tune in to shows that report the types of stories that have a tendency to be those that are paid for, it wouldn't be an issue. But it does. And as long as they do, there are two things that are certain: Paying for the news will always be part of the news shows' budgets, and they will always say it's not.

4

And Then There Was the Truth

In the year 2000, my career was thriving. I was landing some of the biggest news stories around, some of which just fell in my lap. The networks loved me because I gave them stories to fill in the spaces in their shows. It seemed as if I could do no wrong. I was working as hard as ever, but also living that old saying, "Work begets work."

I position a news piece so that it has the highest possible chance to be developed into a book or movie. Some stories require that I act quickly to get them in front of the public, as the attention it receives is over in a flash and any delay would make it worthless. Then there are those that develop over time. This is when patience is required, both in acquiring the news rights and in how the story is released to the media. With this type of story, I earn the subject's trust so that when it is time to acquire the news rights, it is an easy sale. Sometimes, when the press is already all over a subject, I will call them and introduce myself, and give them my name, background, and number, knowing that there is a better-than-even chance that they will be calling me back later to see if there is a book or movie deal. I am not overzealous, because that can scare people away. I'd rather not work with people who don't trust me, and I let them make up their minds on their own.

In early 2000, I was contacted by an attorney who wanted

me to help get word out about one of his clients. At the time, I didn't know it would turn out to be one of my biggest stories, touching almost every soul in the country and reaching all the way to the White House.

George Felos was a lawyer working out of Dunedin, Florida, when he called and asked me to help him get some press coverage for his client. His client, Michael Schiavo, was fighting to have his wife's wishes honored—to let her die in peace.

George explained that Terri Schiavo had suffered a heart attack ten years earlier. Doctors were able to save her life, but not before severe brain damage occurred, leaving her in a "persistent vegetative state." The heart attack was blamed on an improperly diagnosed potassium deficiency, which was speculated to have been caused by an eating disorder. The misdiagnosis led to a substantial malpractice settlement that was used for her care and attempted rehabilitation. George listed the numerous types of therapies Michael had tried over the years to help improve his wife's condition, all of which had failed. It was the opinion of the medical personnel involved with Terri that her situation would not improve. She would be dependant on feeding tubes and constant assistance to remain alive for the rest of her life.

Michael Schiavo was in a court battle to have Terri's feeding tube removed, claiming that it was her wish to die with dignity. It was his position that she would not want to live in that manner, and that he and his wife had discussed the issue after seeing similar situations on TV shows. Terri's parents opposed Michael, saying that Terri would not want to die even in the state she was in. They were fighting for legal custody of their daughter, claiming that Michael had ulterior motives.

In the first conversation, I had mixed feelings both professionally and personally. As far as the media was concerned, I was not sure what to expect. I remembered the case of Karen Quinlan, a twenty-one-year-old woman who had collapsed after drinking too much alcohol mixed with Valium and aspirin in 1975. She was also left in a persistent vegetative state. Karen Quinlan's case received a tremendous amount of media coverage, but she was the first real modern icon in the right-to-die debate. I have learned over the span of my career that when a story comes in second, it receives very little attention from the media. Although it had been over twenty years since the Quinlan case, I was not sure if the media attention would repeat itself.

Personally, I was not sure what to think. Through subsequent conversations with George and some investigation on my own, I learned that Michael was living and had started a family with another woman. I have to admit, it bothered me. Reading some of the statements made by Terri's parents was also difficult. I have three children I love dearly, and I could not imagine the pain they must have been going through. I also felt for Michael. I watched my mother suffer and wilt to nothing as she died. I wanted her to have peace and would have done anything to help her have dignity in her death. I saw news reports that made inferences that Michael's attempts to improve Terri's condition through treatments like brain implants were really attempts to kill her and cover up things he had done wrong. My heart went out to him. I couldn't imagine the pain reports like that would inflict. I also did not know if they were true.

I called George Felos and told him that I would do what I

could to get his case some media attention. I remember saying to George, "We both know that there are two sides to every story, and then there is the truth. Without you being heard it would be impossible for anyone to form a true opinion of this case." I knew that media spin had the potential to tear Michael Schiavo to pieces.

George and I talked on several occasions, but the e-mails from his office were relentless. It seemed like almost every morning for a while I received a barrage of requests and news releases from him. Our work did pay off, and over the next couple of years the case started to get national attention. That is when I started to see the rest of the story, the parts that George had failed to mention.

The lawyer confirmed that Terri had suffered a heart attack and that her marriage to Michael was good. As the case received more attention, different facts started to surface. First of all, there were reports that Michael was living in an adulterous relationship while his incapacitated wife lay hospitalized. Some of Terri's friends had come forward and said that she had come to work with bruises that led them to believe that Michael was abusing her. Some of her friends went so far as to say that Terri was considering divorcing Michael in the months before she became incapacitated. In September of 2002, Dr. William Hammesfahr issued a report at the request of the Second District Court of Appeals. On the first page of his report it said, "In the Emergency Room, a possible diagnosis of heart attack was briefly entertained, but then dismissed after blood chemistries and several EKGs did not show evidence of a heart attack."

Now I was not sure who was spinning who. Was the press

twisting things around to add to the ratings, or was the Michael Schiavo legal camp spinning things to distort the truth?

I found several other disturbing revelations in the course of digging through all the news reports. Terri had a complete x-ray, which showed she had suffered several fractures—including ones to her leg and ribs—but it was not possible to determine the exact time of her injuries. One doctor was quoted as saying, "This girl has been beaten up a lot." The thing I found most disturbing was that her skull had been fractured and there were reports of bruising on her neck, as if someone had tried to strangle her. It was speculated that the injury to Terri's skull could have been the cause of the vegetative state. Michael said, "It was from her falling down, and her bones became brittle when lying in her state later on."

I continued working press releases for Michael's legal team, to get his side of the story out. I was having doubts about the case, though, and found it troubling that the news releases we were getting out were receiving a great deal of airplay on the major news shows. The reports that shed a negative light on Michael were so underplayed that they were easy to miss if you were not looking. It was hard for me to remember sometimes that I had a job to do, to put the facts out there and then let everyone decide for themselves. It was my job to get the information to the media, not tell them how to report it.

In October of 2003, Michael Schiavo and George Felos appeared on *Larry King Live*. I remember looking forward to the interview and seeing how Michael would react to some of the tougher questions, but they didn't pack a punch. The guests had plenty of opportunity to make their points, to include how much Michael loved Terri and what led up to the falling-out with his in-laws, but there were really no revelations. Larry King has my highest respect for journalistic integrity, in fact I have been a guest on his show, but I felt he might have been held back. The ratings game can dictate what questions can be asked of a guest in an interview. If a subject is controversial and the guests can be put in a tough position by granting an interview with no limits on questions, they can decline the interview. That would not be good for ratings. So in many cases, the shows agree to limit questions so the subject will appear on-air. Then the ratings go up, and everyone is happy. I know firsthand how this works, as I have made such interview arrangements for many of my clients.

Whether Larry King held back or not was of no consequence. The Michael Schiavo team got a chance to put on their best case, and Terri's parents' team got their chance the next day. The media monster was happy because more people watched the shows that displayed the family feud. It became clear in that interview that Michael wanted nothing to do with Terri's parents after he received the money from the malpractice settlement in 1997. Both sides said that the relationship was good before that point.

Things quieted down until early 2005, when a Florida court ordered that Terri's feeding tube be removed and her parents had exhausted all legal remedies to prevent it. The Bush brothers, president of the United States and Florida governor, both became involved publicly as the case went to the United States Supreme Court. I was still not really sure what to think of the people I was representing in the media, and their story was the headline on every major news show for weeks. One day, a court would order the feeding tube removed, then another legal move, and the removal would be delayed. After years of legal battles, Terri's feeding tube was removed on March 18, 2005, and America went on a deathwatch. The media monster went into full swing reporting on every court move, civil demonstration, and event that had anything to do with the case. Even Mel Gibson got involved in the effort to save Terri's life. But on the morning of March 31, 2005, Terri died. Her death did not stop the media blitz on her story; it lasted until Pope John Paul II passed away three days later.

But that is not the end of the story.

Along with the reports of Terri's death came reports that Michael would not allow her parents to be at her side when she died, and that he refused to let them have a funeral for their daughter. After five years of working with his legal team, I was disgusted, but still did not want to judge anyone. I hoped there might have been more to the story than I was aware of.

On Friday, April 25, I got a call from George Felos. He had told me that "Michael wanted to meet with me to discuss the future possibilities of a book or movie deal, and he wanted the meeting to be kept a secret." I remember talking to my friend about the conversation. I could not help but to be excited at the possibility of getting the rights to the story of the year, but I also had my reservations. Still, I had to go. My kids were even impacted by the story and begged me to follow up. I am not sure if it was professional drive or curiosity that urged me forward. Either way, I was on a red-eye to Florida that Sunday and arrived early Monday morning.

After checking into the hotel and going over my notes, I decided to get to the meeting early. I planned to be at our 5:30 p.m. meeting at 4:45. It was a forty-mile cab ride from Tampa to Felos's office in Dunedin, and along the way I was a pile of nerves. I had never spoken with Michael before, only to his attorneys, and I had only the Larry King interview to form an opinion as to what he was like. I remember thinking that they had asked me to come out, so I was sure that they were interested in having me put together a book and movie deal for them. Part of me was excited; I really wanted this "get."

When I walked up to George's office, I was a bit surprised. After all the news coverage and conversations over the years, I had a preconceived image that his offices would be like other big-shot lawyers—professional. On the front door to his practice was a piece of paper taped to the door. It looked like it had been taken from the bin of the copy machine and it read "Felos and Felos." As I walked into the office, George's secretary, Francine Wolfe, quickly ushered me back out to the parking lot and handed me her headshot and resume. She made it quite clear that she wanted to play herself in any movie that we produced.

The modest office seemed to be hustling as George came out to greet me, and I felt maybe I had made a mistake by coming early. I told George, "I will grab a quick bite so you can finish up your work." In an instant he asked, "Aren't you going to take us to dinner?" I told him that I would take them to dinner, and with that George asked if I minded sitting in his waiting room for a short time while he did an MSN interview. He came back briefly to explain he had a couple more small things to wrap up, and then had me wait for almost two hours. My time was passed with Francine repeating her limited acting resume and bragging about the ritzy steak house where she had made reservations for us earlier that day.

It was about half past six when the door opened and George, Deborah Bushnell (George's associate and Michael's co-counsel, she made sure to tell me twenty times over the next two hours), Michael, and his older brother Brian walked into the room. I was taken back a little by the size of the Schiavo brothers. I stand about five feet nine, but they both cleared well above six feet. Brian seemed nice enough, but Michael was the one I came to see. My first impression of him was that he was uptight. We made our curt introductions, I was ushered to Brian's SUV, and we drove to a restaurant in the middle of the small town. Brian was so flustered that he pulled out with the back hatch door open and had to get out and close it. I tried to figure out why he was so apprehensive and chalked it up to what the family had been through over the last couple of months.

As we walked into the so-called elegant steakhouse, I could tell that it was going to be a pricey night by looking at the menu. I do not mean to sound petty, but I flew twenty-five hundred miles to talk business, not to wine and dine. The manager of the restaurant was courteous and he seated us in a small, semi-private room, which was out of the main dining area, but still easily visible to the entire restaurant. I had a feeling he knew exactly why I was there and he was astute enough to make sure he provided privacy, as best as he could.

Besides a little small talk, very little had been said by Michael until he sat down to the table. As he leaned back into his chair he looked at me and said, "This is my first night out, I am grieving my wife's death," in a voice just shy of sincerity. I felt like asking what his fiancée and children were doing, but thought better of it. I also wanted to get a feel of where he was coming from. I am not sure how I would have reacted with all of the media attention he had received over the last few weeks, added to Terri's death. It was clear he had been through a lot, so I decided to just let the evening play out and see what happened. In my heart, I believed in the story and felt it could be important. I did not want any of my preconceived notions to tarnish the chance of something good happening with the piece.

We had been seated for a few minutes and were looking over the menus when the waitress brought over a huge plate of appetizers. When Michael asked whom they were from, I could feel his tension hit the air. I asked, "Do you think they might be from someone who disagrees with your position?" He shrugged it off and said "No," but he started to eye people who where walking by. I then asked, "Would you like to change chairs with me so your back will be to the public," to which he replied, "Look, strangers have made me." At that point the tension was coming from within me, and, as cliché as it sounds, I felt like I was in the middle of a *Godfather* movie and I wished my back were not to the door.

The waitress took our orders, the brothers sipped on their drinks, and George finally started to talk about why I was there. "Why should we go with you?" he asked. I was shocked and angered. I thought about his question for a moment and replied, "After five years of getting your side of the story out to the press and maintaining a professional relationship, I think that would be a good *why.*"

I was being auditioned like a kid fresh out of acting school. It had been years since I'd had to sell myself, and it was especially unsettling to be treated that way by people I had worked with for so long; but I let it slide. I wanted to see what they had on their minds. I knew an endless number of people were most likely approaching them for a book deal. With that and the timing of everything, I decided to let them have control of the conversation and see what they were looking for. It didn't take long to find out.

There was a little more small talk, and then the question was asked: "Do you think a $5-million advance for the book is realistic?" George threw it out on the table as nonchalantly as he could. It was all I could do not to laugh out loud. Advances are based primarily on anticipated revenues; to earn $5 million the book would have to sell more than 2 million copies. This was a big story, sure, but not that big. Michael and Brian joined the conversation, telling me about the hoards of literary agents that were throwing out incredible numbers. I felt like the only thing being thrown around was a bunch of BS. All they wanted to see was how much they could get me to pay them upfront for their book and movie rights. I let them talk about all of the so-called offers and endured them saying that they had no idea who I was, mixed in with the occasional, "Why should we trust you?"

It was time for me to bring them back down to earth. I explained that the story they had to offer was important, but would soon be a memory that would fade from the public's mind. I told them that Monica Lewinsky's book deal was for $1 million and that what they had did not hold a candle to what she had as far as the media was concerned. I also had to explain that not everyone in the country liked Michael, in fact a lot of people didn't, and that wasn't going to help to sell more books. I thought it was odd that when I said that, Michael smiled. In short, I tried to give them a condensed education on how the system worked, but it fell on deaf ears.

George and Michael were silent for a few moments, and I thought maybe I had made my point. Michael then looked at me and said, "I was offered ten million to walk away from Terri and let her parents take over her care and custody. The money was coming from a group that was against the right-to-die movement. I turned them down flat." His eyes were locked on to mine and I felt that he was trying to control my thought process with his glare. I remembered seeing the reports about the offer, but wasn't sure I believed it. Two seconds later he blurted out, "So, how much can you get for me upfront, even if you have to put it up yourself?" It was just plain weird. In the same breath, he wanted me to believe that his convictions did not have a price tag, yet he wanted an unreasonable amount of front money for his book deal. I couldn't help but feel that the offer was bogus and Michael thought it would help him better his position to get more front money; or perhaps the Schiavo team thought there was a bigger paycheck, in the high eight figures, in letting the story play out with Michael in control of Terri to the end.

A few minutes before the meal was served, Michael looked up and said, "See, they are looking at me again." This time George suggested that he trade seats with Michael, but Michael would not hear of it. Then Michael burst into tears. I thought the guy was either manic-depressive or a very talented actor. Michael's brother turned and looked me in the eye and said, "If anyone hurts my brother, I will kill them." The environment at the table had changed so quickly, so dramatically, that I started to wonder if this story was really worth it. I do not remember ever feeling so uncomfortable.

The rest of the night passed slowly. Following the appetizers, which my dinner companions finished off the before I could touch them, and our meals, Michael and Brian ordered and quickly downed after-dinner drinks. It seemed that Michael didn't hold his liquor very well. He became even more assertive in his demands for front money.

There were a couple of times when the lawyers left the table to use the restrooms that I thought I might have the opportunity to really get a feel for Michael's mindset. It seemed odd to be sitting with a man who was the contradiction of the moment. He was either the most hated man in the country for pushing to have his wife die or a hero for letting her die. I wanted the truth, so I sent out a feeler. "So, Terri would probably have loved a movie or a book done on her?" I asked. "Bull, she hated publicity and would have never wanted any of this crap!" he answered sternly. "I don't understand what those scummy Hollywood producers would want with my life." I was at a loss trying to figure out why he and his legal team had me fly out, but I tried anyway. I looked him straight in the eye and said, "I guess then you want to help the right-to-die movement and other people?" Michael then looked down at the table and said nothing.

I felt sick. The evening was not going very well. I decided to try and salvage the situation and find some of the answers they were going to have to address, no matter who did their book or film. I just wanted to see if I helped the wrong person. I had read about a death threat made to Brian's wife and asked him about it. I was surprised when George looked up and claimed he had never heard about the incident. Michael told his lawyer that it was his other brother, and he had neglected to tell him about it. *Strange,* was all I could think at the time. The conversation switched to the account of one of Michael's friends who lived in New Jersey. The story was that he was out one night and saw a bumper sticker that read, "DON'T FEED MICHAEL SCHIAVO," which led to a fistfight between Michael's friend and the owner of the car.

In my heart, I knew it didn't matter if they wanted to do a book and film with me or not. If Michael couldn't convince me that he was telling the truth over the years about Terri and their relationship, I wouldn't do it. The story could be important, but it had to be true. I asked about the eating disorder and he responded only by saying, "Her potassium level was low." It seemed odd that he couldn't or wouldn't elaborate on the issue that made the court rule in his favor in the malpractice suit. As much as I wanted more of an answer, the tension was too high to push the issue. I then asked about the discovery that Terri had suffered broken bones, and I could tell that made George uncomfortable, because he quickly mentioned that her bones had become brittle because of her lying in bed all those years. It was as if he was afraid of the logical question: "Did Michael abuse his wife?"

As I asked them to address some of the more controversial issues that any book or film would have to address, they quickly changed the subject back to what was in it for them. The only thing they wanted to hear about was money, my credibility, and how I was going to prove myself to them. I called my agent from my cell, who repeated what I had already told them. Bill Gladstone is the president of Waterside Entertainment and a very successful agent. He also did other book deals for me and I trust Bill implicitly. He told them candidly, "No, your book rights are not worth five million dollars;

and yes, you will have to address all of the issues, not just the ones you want to talk about." I then called Kent Walker, a client I had represented in the past (and my collaborator on this book), and had the Schiavo team talk to him. Kent told them that I had put together successful book and movie deals for him. In short, I had provided everything they had asked for to show my credibility, but they were not finished.

"I want total creative control and to be able to interview any of the actors when this thing goes to a film," Michael blurted out. I remember staring in disbelief. Five million dollars up front; tell the people in Hollywood, who make movies for a living, how to do their jobs; and have me represent them. I was starting to wonder what the hell I was doing there and hoping I could get an earlier flight home. I took one last shot at salvaging the evening. "Look, you need to come down to earth. I am not a studio who is going to give you carte blanche. I am a producer who wants your story to be portrayed in a positive and honest manner, or not at all." And with that, the deal was dead. They weren't interested in hearing about the moral issues of the story, just whether or not I would be the highest bidder. I could see in their eyes that my trip was a waste and all I would have to show for it was a credit-card statement with a few thousand bucks worth of airfare, hotel, and an expensive dinner to show for my time.

I gave up. At that moment I decided that I was not disappointed that this "get" would pass me by. I felt relieved and cursed myself a little for not listening to my gut over the previous five years. They ordered and finished off their deserts and after-dinner drinks, and I didn't eat a thing. In fact, I remember hoping that room service kept late hours at my hotel, because I knew I would be hungry when I returned. That is, if I still had an appetite after thinking about Terri and her family on the cab ride back.

As we got up to leave, I paid the bill and reached for my camera that I had brought for the occasion, thinking at least I would have another picture for the wall in my war room back home. They refused the photo op, saying that they wanted to keep our meeting a "secret." I smiled as we stood in the middle of the crowded restaurant, thinking that someone had sent over a plate of appetizers after recognizing Michael and George a little over an hour before. To this day, the part of my wall I had reserved for this story remains empty.

I heard them say as we were leaving that they were going back to George's office, and I mentioned that I could catch a cab from there. "Oh, you can catch a cab from right here," George said, and Michael nodded in agreement. As they said good-bye, they acted as if there were done with me and wanted nothing to do with my company anymore. There was not even a hint of appreciation for dinner. As they walked away, Brian turned around and put his hand on my arm and said, "I would like to apologize for the way they acted, but I want you to know that I am the one who makes the decisions for my brother." I thanked him and walked into the bar to call a cab. While I waited, the manager of the restaurant came up to me and said, "We heard it all, and it was only a matter of time before someone did a movie and book on him." I smiled as he handed me his card saying, "My name is Jimmy Stewart, and I would love to be in the film." I was trapped in nowhere-land surrounded by people swimming in delusions of grandeur, who thought they could be considered for a movie role by handing me a business card.

A guy who wanted to give me his two-cents worth about Michael Schiavo and his story greeted me outside, but I had had enough. I just wanted to get back to the hotel, so I made a beeline to the waiting cab. As I climbed into the backseat of the taxi, I saw a woman seated next to the driver and thought it a little odd. I soon learned that the woman was the driver's wife, and she had come along to keep her husband company for the shift. It was actually kind of cute; she gave her husband directions and they made small talk about the town of Dunedin. We drove by the town's barbershop; it reminded me of the one in Mayberry on *The Andy Griffith Show*. They bragged about a stone-throwing contest that had taken place the day before as we passed by City Hall, which looked like a model I bought my son years ago for his train set.

They were such pleasant people, talking about fishing and Dunedin's Scottish history. Their company was welcome relief after the dinner I had just endured. After a few minutes, though, the conversation turned to Dunedin's most famous citizen. "You heard of Michael Schiavo? Well, he lives here. He is a nurse like me," the woman said. She went on to say that her friends worked with Michael, but didn't really like him much. "They said he is very overcontrolling and has a problem with his temper," she reported. I sank into the backseat and resigned myself to the fact that I was not going to be able to escape the subject that night. I also began to understand more clearly why I had that bad feeling in my gut about Michael.

After years of seeing both sides of the story and trying to remain objective, I finally knew what to think about the situation. Yes, the Schiavo team was money-hungry and used me to position them for a book and movie deal. It was obvious they had money on their minds and they wanted to milk the situation for all they could get. I deal with people like that all the time. What made this situation different was that it was also obvious to me that they had been thinking about a media payoff for years, not just in the weeks following Terri's death. I also know that I have never been so happy to hear the wheels hit the runway as I was the next day when my plane landed at LAX.

There was some correspondence over the next couple of days. I wrote a straight-to-the-point letter to George explaining how the system worked and how he was not going to change it in answer to his e-mail stating that Michael needed a little time to make his decision. I also sent over my credentials. It was important to me for them to know they had been working with a pro, even though I didn't want anything to do with the project. I guess you could say that my pride got the better of me.

The story is too important to be petty. But, after the experience I had to endure in Florida, I had more questions than answers about Michael Schiavo and his legal team. I am a businessman, but there are more important aspects to a story than just how much money can be made. I need to be sure that a project is honest. I also need to feel secure in the sincerity of the people I work with; and honestly, I really had to doubt the Schiavos.

A few days after I returned home, a retired nurse who had worked in a Florida hospital contacted me. She claimed to be friendly with a wealthy man who owned a pharmaceutical company. According to her, the man said he knew the people who made the so-called offer of ten million to have Michael relinquish control of Terri's life. So, maybe the offer was true, but why then was Michael so money-hungry when we talked about our deal at our dinner? It even crossed my mind that maybe Michael started the rumor. After seeing how he acted at dinner, I would not put it past him. I also wondered about Michael's actions after Terri's death. He did arrange for an autopsy, which appeared to be under his control, but then had Terri's body cremated quickly after a nonreligious service. Cremations are highly unusual for those who are Catholic, as Terri was.

If the opportunity had presented itself before our dinner, I might have taken the story to develop into a book and movie. I have been in a business where spin is a way of life, and I guess I tried to spin myself into believing that Michael's intentions could have been sincere.

A couple of weeks later, my curiosity got the best of me, and my gut feelings were further vindicated. On May 9, 2005, I sent the Felos office an e-mail:

> Just curious as to what is happening?
> Larry

I had to laugh out loud when I read the response after my computer chimed, "You've got mail!" a few minutes later. It caught me a little by surprise:

> Larry:
>
> Glad you asked. I was going to e-mail you today and had not gotten around to it yet.
>
> Mike told us he has decided to negotiate book and movie deals himself. He told us to refer all inquires to Brian's cell phone. Here's the number: [No, I am not going to give you

> the number]. You may want to contact Brian right away to start direct talks if you are still interested.
>
> Sorry we won't be able to work together on this, but I enjoyed my phone and e-mail correspondence with you, I hope you and Mike/Brian are able to work something out.
>
> Debbie

I picked up the phone and called Brian. The first thing out of his mouth was the confirmation of Debbie's e-mail. Michael wanted to take care of his own negotiations regarding book and movie deals. It was also easy to see that they wanted to keep their options open. "I'm sorry for the way you were treated when you came out here. Michael and I would have driven you to your hotel, and we did not understand George's [Felos] behavior," he said, apparently forgetting his brother's behavior. Brian also claimed that Michael was not aware that I had been the one to get the press they wanted over the previous five years.

After the conversation was over, I became convinced that Michael's devotion to Terri's right to die was motivated by more than just her wishes. It became too easy to see the greed. George Felos represented Michael well for years, and when it came to a movie and book deal, I'm sure George did his homework. I can almost picture the two of them following our dinner together, sitting at a desk with Brian next to them as George explained that what I had told them was true. Then I can see Michael not wanting to see the truth and firing the lawyer that helped make him newsworthy in the first place.

If someone wanted to be a skeptic, this case would make it very easy. Did the events show a loving husband's devotion to follow his beloved wife's wishes? Or, did the events appear that someone was trying to cover something up and then profit from it?

I try to be objective with every story, but if someone were to ask me if the allegations that have been brought against Terri's husband in the press are true or not, I couldn't swear to anything. All I can do is tell you about my experience with Michael Schiavo. As far as what his motives were, or are, you will have to judge for yourself.

5

Hush Is an Understatement

The horrific events of April 19, 1995, could have unfolded in any town in America. The workday was just beginning as people rushed to Starbucks to get the caffeine jolt they needed to get started. Others dropped off their children at babysitters. In the minutes before 9:00 a.m., the day was just getting into gear. The streets were crowded with the hustle and bustle of people trying to get to their respective offices on time. April 19, 1995, was a pretty spring morning in downtown Oklahoma City; and just like any other day in any other large city, the routine was being played out. No one gave a second thought to the large yellow Ryder rental truck parked at 200 NW 5th Street. Rental trucks were always making deliveries and there was nothing to indicate that this one was any different, even though the driver left the truck and was out of sight within minutes of parking it.

That was until approximately 9:02 a.m., when a thunderous explosion fueled by ammonium nitrate and nitro methane—a mixture of agricultural fertilizer and volatile racing fuel also known as ANFO—roared through the downtown area. Windows shattered and walls shook as far as blocks away. Those who were unfortunate enough to be in close proximity to the rental-truck-turned-bomb died instantly, as the concussion of the blast crushed the vital organs of their bodies just before blowing them to pieces and vaporizing the remains. Those further from the center of the blast burned. Their flesh was charred as panic forced rapid breathing of the scorching air to burn their lungs, a quick but painful death.

About thirty feet from the truck was the Alfred P. Murrah Federal building, an office complex that housed the federal offices in the Oklahoma district. The people who were located on the side of the building opposite the explosion had a chance, as the building itself acted as a shield; but for those on the side of the blast, it was a different story. The front half of the nine-story building took the brunt of the force caused by the explosion and collapsed within seconds, trapping people who worked in the building under tons of concrete and twisted metal. What had been a nice-looking office complex moments before became a pile of rubble. By the time all the dust settled and the rescue efforts had been exhausted, 168 people, 19 of whom were children, had lost their lives.

The news media was in full gear within minutes of the explosion. Major networks interrupted the regularly scheduled programs for special reports, and for the rest of the day just about every television set in the country was displaying images of the shattered building and the victims. People could be seen standing in the shells of offices that had been sheared in half many stories from the ground. Some media outlets jumped the gun and reported that it was a terrorist attack. Then-President Clinton went on-air hours later, asking the media and the American public to be patient and not to make any assumptions as to who was responsible. He promised that the authorities would find out soon, and he came across very confidently, as if he already knew.

About ninety minutes after the Murrah Federal Building was blown almost in half, twenty-seven-year-old Timothy McVeigh was stopped and arrested by an Oklahoma Highway Patrol officer for driving without a license plate. He was about to be released from custody two days later when he was recognized as a suspect in the bombing. McVeigh had an old buddy from his army days named Terry Nichols who was also a suspect. Nichols surrendered to the authorities in Herington, Kansas. Both were soon charged with the bombing and the media monster had its story for months.

The story of two old army buddies conspiring and carrying out the worst terrorist attack in U.S. history, at the time, seemed to fill the American public's need to know. The why was said to be a vengeful attitude about the handling of the incident in Waco, Texas, which began on February 28, 1993, when the ATF attempted to arrest David Koresh at his ranch for possession of illegal firearms and explosive charges. The siege ended April 19, exactly two years to the day before the Oklahoma bombing, with the death of Koresh and seventy-six of his followers. Among the dead were twenty children. The Waco incident was still fairly fresh on the minds of the American public when the Oklahoma City bombing occurred, as the media had covered it extensively, including the fiery conclusion. Although horrified that such a random act of violence of this scale could happen on U.S. soil, everyone seemed to be satisfied with the answers that were presented by the media.

It seemed to be too simple an explanation, and my suspicions only grew as I began to procure the news rights from some of the people who became part of the tragic story. I spoke with Lana Padilla, the ex-wife of Terry Nichols, and their son

Josh, a few weeks after the bombing. At that time, they were completely shocked that Terry had anything to do with the deaths of the 168, although over time Lana has worked through her denial of what happened and now sees how he was involved. I also procured news rights to the story of Kathy Wilburn and her daughter, Edie Smith. Edie lost her two sons when the Murrah Federal Building collapsed. At the time, I was working with Time Warner and was instrumental in getting her image on the cover of *People* magazine, bringing more attention to the tragedy that took place in the lives of those who lost loved ones in the disaster.

Kathy was proactive in finding answers to the events leading up to the morning of April 19, 1995. With her information, and putting together some of the pieces I received from other sources, including what Lana and her son had told me, I became aware of parts of the story never mentioned on the news.

Nichols and McVeigh were not alone in their hatred of the ATF's actions in Waco, Texas. In fact, they had a kind of club, or gang if you will, that shared their feelings on the government. From the outside, it might have looked like a white-supremacist group, but on the inside it was pure hatred for the American government. According to a few of my sources, the FBI was aware of this group months before the bombing in Oklahoma and had even infiltrated the group with an undercover agent.

What I found to be most disturbing was that a couple of my sources told me that the FBI had knowledge of the plans for the bombing days before it took place. Equally upsetting was the report from one person who told me that the nine-member ATF office housed in the Murrah building was not occupied during the explosion, although they had no proof to offer.

It was frustrating that the news media did not want to even entertain the idea that something other than what the government officials were dealing out might be newsworthy. The media was very receptive to the survivor stories that I brought them, and they received a great deal of airtime. But when I brought them evidence of the possibility that the disaster could have been completely avoided, they closed their doors. As a newsperson, I know that not all the leads and accounts of a story are going to be accurate. But I also know that you have to follow up on every piece of information if the truth really is all that matters, even if you do not like what the truth is going to look like. I made several attempts to get the attention of the news directors, but it quickly became clear that they would have nothing to do with alternative theories.

To this day, I can't swear whether or not the possible conclusions I tried to bring to network news were accurate. I do know that the information I had was reliable, but it was only part of the whole story. Recently, Lana and Josh Nichols got information from Terry Nichols and have signed with me to do a new book. They are in the process of supplying me with that "Oh, my God" information that everyone is dying to know.

The press does have the ability to find out just about anything on anybody or any event, if they are so inclined. For some reason, they were not inclined to follow up on the very real leads I tried to give them. I have ample reason to believe that the public never did hear the entire story on the bombing, and that the higher-ups in the news agencies went out of their way not to report on specifics that the American public should have known.

I have spent the last twenty-five years of my life working in and around the news industry, and the one thing I have learned is that television news has a power never before seen in history, and I very much doubt that it will ever have an equal. Cameras can put you in the middle of a story as it unfolds—that is the base of their power. But we still have to rely on the facts, or at least what we hope are the facts, that the reporters tell us as we watch the video. Then we have to hope that once the video stops and the rest of the news is told from an anchor's chair, what we hear is the truth and that all of the truth is being told. In a democracy, the news media has a tremendous responsibility as it is one of the checks that distinguishes a free society from a dictatorship. The Founding Fathers realized this, enough so that the First Amendment was drafted. It is not as simple as it sounds though, and when we look at the evolution of how the news is brought to the masses, we can see how the people who bring you the news have a responsibility and a power that only grows.

Each of the last few generations has had those moments that hit the inner core of each individual, now more so than before, because our technology-driven media can now focus our collective attention on one single event in a moment of time. When people relied only on newspapers for information on what was happening in the world, it could take days or even weeks to find out about something important. When Booth shot Abraham Lincoln, it took over a month in many cases for people to hear of the assassination, and even then the details were prone to be sketchy at best. The world really was bigger.

Then came the wires. As daily events manifested, people around the country could find out about a major event relatively quickly, as long as they happened to be near a telegraph station or newspaper outlet or the place where the dramatic event occurred.

Broadcast radio changed things forever. From the time the majority of people in the country had live broadcast playing in their homes, and the farmer in the Midwest could hear the same voice at the same time as the stockbroker in the east, as the fisherman in the northeast, as the cowboy in the southwest, we all could be bound by a news story like never before. On May 6, 1937, America got its first taste of living a tragedy as it happened, as a nation, when the German zeppelin *Hindenburg* exploded above a field in Lakehurst, New Jersey. Everybody listening to the live broadcast felt the horror as the radio announcer described the unexpected inferno that resulted in the death of thirty-five people. I remember as a young boy my mother taking out a small piece of cloth to show her guest; it was a part of the famous airship's hull.

Television amplified the progress of radio and the effect tenfold, maybe a hundred-fold. With television, not only can we hear a story, we can watch as it unfolds, and the images we watch on the small screen in our living room will grab us even deeper than words alone. And television has never released its grip.

Someone in New York can watch a tragedy on a TV screen at the same time as someone in California, and the shared emotions no longer are confined by time or geographical boundaries. Viewers don't have to use their imaginations to picture what a scene looks like; the pictures on the tube take care of that for us. Seeing an event in your mind's eye is one thing, but watching the actual event is quite another. Because of television's ability to cross boundaries that radio never could, television binds a nation and sometimes a world as tightly as a small village. These moments can be recalled years later, and the same feeling of that instant is felt again in the pits of our stomachs. Radio had its moment on December 7, 1941, "a date which will live in infamy," when Pearl Harbor was attacked, throwing America into World War II. That generation could only hear about the tragedy as it happened. There was also a risk of misunderstanding inherent in only being able to hear the news; like the night that many Americans were convinced Martians were attacking, when Orson Wells broadcast his radio play *War of the Worlds* on October 30, 1938. Television has brought our generation several of those moments in live pictures, leaving very little room for misunderstanding.

November 22, 1963: A shot from a book depository killed American president John F. Kennedy and ended the modern-day Camelot.

July 21, 1969: Neil Armstrong took "one small step for man and one giant leap for mankind" when he stepped onto the surface of the Moon.

August 4, 1974: President Richard Nixon resigned his office.

January 28, 1986: The Space Shuttle *Challenger* exploded, killing all seven astronauts onboard.

April 19, 1995: The Alfred P. Murrah Federal Building in Oklahoma City is bombed, killing 168 people, the worst terrorist attack disaster in American history at the time.

September 11, 2001: The World Trade Center fell to the violence of a terrorist attack; a hijacked plane, thought to be headed for the U.S. Capitol, crashed into a field outside of Shanksville, Pennsylvania; and the Pentagon was directly attacked by a terrorist-hijacked passenger jet.

Each of these moments has something in common. Most adults remember exactly where they were and what they were doing when they first heard of the events. Walking on the moon was a victory for every human on the planet; a president quitting his post shook a nation. But most of the other events were tragic, and some have had far-reaching effects on the lives we all live. Television news and the pictures they broadcast are the way these stories reach deep inside of you. Unfortunately, you can't always believe what you see, and you cannot see everything. That is where the trust of the media comes in. We have to be able to believe that all relevant parts of the major stories are being told. Being part of the democracy that freedom of speech is designed to protect, we shouldn't have to second-guess the information brought to us on TV news; but we do.

The thundering crash of twisted metal collapsing onto itself with the silent screams of more than three thousand men, woman, and children losing their lives will echo forever. Passenger jets fell from the sky and crashed into fields, the Pentagon, and the World Trade Center on September 11, 2001. In a moment, the world was changed forever. Few events have had such a direct and profound impact on our day-to-day lives as a nation as did the 9/11 attacks. There have always been and always will be big news stories, but nothing has captured the attention of the nation and touched us all so deeply as the day America realized that our oceans were no longer enough to protect us from hostile forces. It changed the way we conduct ourselves as a nation, and the news industry was not immune.

Sensational stories that captured the headlines in the days before 9/11 weren't so important in the months that followed the attacks. For a while, people didn't care about the wrongdoings of a congressman, even if they might have included murder. Do you remember Rep. Gary Condit's story that captured the headlines in the months before 9/11? His intern, Chandra Levy, went missing on May 1, 2001, and in the course of the police investigation it was rumored that the two were having an affair. Condit was on the defensive. A nation saw him as a trusted public official who was guilty of the murder of his intern/girlfriend, and it would be hard to find a tabloid, magazine, or news show that didn't feature the story. After 9/11, people didn't even remember his name. We didn't have time to update ourselves on the naughty things the celebrities were up to, and we really didn't care. Our priorities changed when we realized the world was not as safe as we had once assumed it was, and we were all scared. Some of us a little, some of us a lot.

I remember the exact moment when I learned of the attacks on 9/11. I live on the West Coast, so it was three hours earlier than the East Coast when my son called my cell and told me to turn on the news. "Dad, airplanes are crashing all over the place and New York is on fire!" Even though I have been in the business a long time, I was not immune to the fear that the rest of the country felt about our national security as I watched the events unfold on my bedroom TV. The one thing I did know was that I would find answers sooner than most, and I felt obligated to make sure that they knew everything I knew.

In the days, weeks, and months after 9/11, my phones were inundated with calls from everyone; from people who claimed to be responsible to people who claimed to be survivors, but quickly turned out to be fakes. A few months after the attacks I received an e-mail from a man I knew only as Carl. His story haunts me to this day. At first, I thought he was just another quack trying to cash in on the horrible event, but when he told me he had a video of a missile flying into the Pentagon, not a passenger jet, I listened. When he told me he would like to meet me and show me the video, he got my attention. When he told me that the FBI was trying to stop him from showing it to anyone, I got nervous.

Conspiracy buffs are usually a waste of time—but this one was different. I could feel the sense of urgency in his voice, and the sincerity. He claimed to have videotape that would clearly show that the Pentagon had been hit by a military missile, not a commercial airliner. For the most part, my twenty-plus years of experience helps me weed out the fakes; this guy sounded real. In the weeks after his call, I started to look into the possibilities myself. From the day of the attacks, I was bothered that not one picture or video was captured of the jet that flew at a very low altitude near our nation's capital and targeted the center of our country's military might. With the World Trade Center there was a video of the first plane smashing into the North tower and countless pictures of the second plane impacting the South tower. I have been to Washington D.C. and its surrounding areas on many occasions, and I feel very safe in saying that there are more reporters, cameras, and video cameras per capita there than any place in the world. There were several eyewitness accounts, but not one piece of video for the networks to play over and over in the days after the attacks. When the Pentagon video was aired, it was just the aftermath of the explosion. In a world where a video camera can capture a bunch of cops beating up some poor black guy in a back alley, I find it hard to believe that not one camera captured the 757 screaming above a very densely populated area toward the Pentagon—though the government is now showing something that to many does not look like a plane in a recently released video.

A strange phone call a few weeks before Carl e-mailed me almost escaped my mind. The man claimed that he had evidence that it was impossible for the damage at the Pentagon to be caused by a passenger jet because it was impossible for a plane of that size to fly at an altitude that low, at that speed. On one of my business trips a few days later, which happened to be on a 757, I remember talking to one of the pilots before he boarded the plane. I asked him how many hours flying time had he accumulated in his career; it turned out to be thousands. I then asked him if he had the skill to fly a 757 ten to twenty feet off the ground at a speed of over 500 miles an hour, to which he replied, "Chuck Yeager couldn't do that! If it was a performance aircraft, maybe, but planes the size of a passenger jet do not react in an instant. You have to fly ahead of the plane because the control inputs take longer to change the plane's altitude. There are larger surface areas and the sheer weight of a big jet would make it impossible to have the degree of control you would have to have to fly at that altitude without crashing into the ground. I really don't think it would be possible." I knew I was pushing it, but I went ahead and asked him what the chances would be of a novice, with very little training like the hijackers were reported to have, being able to hit a target with pinpoint accuracy. He said, "Zero," then added, "The auto pilot isn't even that good, and if it were on, it would not allow the plane to fly at a low altitude, let alone tree-top level. Someone without training would have a hard time hitting the Grand Canyon."

The Internet buzzed with alternative theories of what had really happened. The major networks even gave a little airtime to some of the theories, but soon swept them under the rug, for reasons unbeknownst to me. Most of the so-called evidence that many bloggers tried to use to convince us of their points seemed weak, but some of it was noteworthy: the impact area of the Pentagon, for one. I remember seeing pictures of other aviation disasters when jetliners had crashed into buildings, and the pictures clearly showed wreckage of the plane. In the pictures that were broadcast of the Pentagon, there was very little of the aircraft shown. The 757 is a large plane, and in my mind a plane weighing 220,000 pounds at take off should have left more than a few pieces of wreckage. And what little wreckage that was found was shown to be painted differently than the American Airlines fleet.

I also found the pictures of the impact site troublesome for a couple of other reasons. One, the grass right in front of the impact point was not burnt at all. In fact, it looked like the landscapers had just cut it, even though a hundred tons of plane and jet fuel had crashed and exploded a few yards away. Two, the hole in the building was not big enough. When an image of a 757 is superimposed on a picture of the impact point of the Pentagon, something is wrong. The jet's wingspan is over 123 feet, the hole isn't—not even close. Even taking into consideration the explanation given by government officials that the wings folded back during the impact, the hole in the side of the building is still too small. Every expert I have spoken with says it is impossible for there not to be any signs of impact points from the twelve-thousand-pound engines on the side of the building. The windows were not even broken where there should have been holes caused by the wings and engines. It looks like a missile hit instead of a plane-crash site.

There's more. A video of the plane crashing into the Pentagon, but no one had ever seen it. The Sheraton Hotel and a service station had security cameras pointed in the right direction to capture the images of what exactly did hit the Pentagon, but the FBI confiscated the film the same day as the attack. The employees were told not to discuss anything they saw on the videos, especially with anyone in the press. I still have a hard time understanding how a video of something that has already happened and been reported on could pose a threat to national security, unless the video was of something different than that which was reported. In any event, after my research, I wanted to get my hands on the video Carl claimed would answer the questions.

Carl could not be convinced to meet with me in person—or even give me his real name for that matter. He did e-mail me a copy of the video a couple of weeks later. The video was not as clear as I would have liked it to be, but it left no doubt whatsoever that what hit the Pentagon on 9/11 wasn't a 757. The image looked like a smaller plane or cruise missile, although the quality made it impossible to know for sure. After a little more thought, however, I realized that it didn't make sense that it was a smaller plane; it wouldn't have put more than a dent in the side of the reinforced walls of the Pentagon. The damage penetrated the five rings of the fortress to the innermost ring, the last to be damaged in the attack. When I look at some of the news archives and compare the damage to the Pentagon to other concrete buildings that have been hit with a cruise missile, I have no doubt in my mind that something other than a 757 hit the Pentagon on 9/11.

Carl wasn't after a book or movie deal. He didn't want to go on the morning talk shows and get his fifteen minutes of fame. In fact, he didn't even want his name mentioned by the media. All he wanted was to get this tape out to the public so the media monster could do the digging and find out what really happened. I saw his intentions as nothing less than patriotic. I had no doubt in my mind that when I brought this video to the networks it would be the top story, and the media machine would kick into gear and mobilize all of its resources to discover, or uncover, what really happened. I was wrong.

Immediately I sent copies of the video to a couple of the major news agencies after brief phone conversations. Each conversation was the same. I explained what I had with the video and the networks said, "Oh, my God! Get that tape over here right away!" After sending the tape over, I sat and waited for the next couple of days watching the news. I couldn't wait to see how this headline was going to play out. Instead of getting to watch the headlines, I got a phone call instead. "Larry, you need to listen to me on this. The video never existed, you never saw it. This could cause some real trouble if you pursue it any further." Sometimes what people say causes an impact; this time it was how it was said that caught my attention. I remember hanging up the phone knowing that I could be putting my family and myself at risk if I tried to push the issue, and I knew there was no way to protect myself. So, I backed off. Being a journalist at heart, I wanted to find out what really happened. Being a dad, I wanted my kids to be safe.

Two things continue to run through my mind in the years since 9/11. One, what really happened to the 757 and its passengers that no one caught on tape or on film, crashing into the Pentagon? Two, this was not the first time I had a tape that seemed contradictory to what was being reported and told that it didn't exist.

On July 17, 1996, a TWA 747 jumbo jet sat on the tarmac at John F. Kennedy Airport in New York waiting for clearance to taxi for takeoff and start its flight carrying 230 people to Paris, France. After long delays, the flight was finally cleared for take-off. Twelve minutes after the wheels left the ground, at an altitude of 13,700 feet, an unexplained explosion killed all passengers and crew. TWA Flight 800 became a household phrase.

It was pre 9/11, and the public was not as, for lack of a better word, comfortable with the term *terrorism,* as it had been years since Pan Am flight 103 met with tragedy over Lockerbie, Scotland. Americans were accustomed to two types of explanations for air disasters: pilot error or mechanical failure. All the major news agencies were scrambling to get the scoop from the FAA as to what caused the crash, and it was a horrible crash at that. Re-creations showed that after the explosion, the plane broke in two at the rear of the signature humpback of the 747. The front piece fell away into the ocean immediately, and, in a freaky play on the law of aerodynamics, the rear of the plane continued to fly for a short time. I cannot imagine the fear of those poor people in the rear section of the plane, as the wind blasted through the cabin at deafening levels, holding onto hope that they might still have a chance to live. They had no way of knowing that the pilots and the controls to the aircraft were already sinking into the sea below. The imbalances and extreme forces finally tore the plane apart in the sky, and bodies, fuel, and wreckage struck the ocean surface at terminal velocity.

The public wanted answers, and the families of the victims wanted them quickly. Over the course of the investigation, it was concluded that while the plane was delayed on the ground, a fuel pump in the center fuel tank had become overheated, or superheated. As the plane was ascending to cruising altitude, with the conditions just right, the overheated pump sparked the fuel vapors and an explosion resulted that tore the jet apart. Even though it took almost a year for the National Transportation Safety Board and the FAA to come to these conclusions, it seemed to settle the public's appetite for the causes of the crash. The FAA published new procedures on fuel management for 747s, especially if they sat at the gate too long, and the public was led to believe it was safe to fly after all. The accident was caused by a one-in-a-million goof.

Recovery efforts to retrieve the bodies of the victims and the wreckage took months. As the wreckage was being reconstructed, public inquiries were held into the cause of the air disaster. Just like years later, after the 9/11 attacks, my phone rang off the hook with people who were either really part of the story or just wanted to be. I interviewed some of the families who lost loved ones on TWA 800 and a few of the eyewitnesses who saw the disaster happen. One thing caught my attention: The public statements the FAA released did not jive with what the eyewitnesses had told me. An internal spark was the FAA's position for the cause of the explosion. I had several people who claimed to see the plane hit by something external before the explosion. In fact, several stories were aired, some of which were from people I interviewed, reporting that many people saw what looked like fireworks fly up from the ocean, then one or two explosions high in the air.

When several people watch a car accident and are later asked to describe what they saw, you will most likely get several different versions of the accident. A little difference here, a little twist of a fact there. But, when you put the whole picture together with all the accounts, you can usually figure out what actually happened. I talked to over twenty different people who saw what happened over the Atlantic Ocean just south of Long Island, New York, on that fateful night. There are differences in the stories. Some said it came from the beach and was bright like a rocket; others said it came from off the coast and was dim, almost hard to see. Half saw one explosion, the other half two. All of the accounts are just a little different, but when you put them all together, the story I get is that TWA Flight 800 was shot down or hit by something. Even with all of the accounts of the eyewitnesses, there was still a lot of room for doubt. That was until about two months before the FAA released its findings on the cause of the accident, when I was contacted by someone who wanted to deliver proof that the 747 was, indeed, shot down by a missile.

When I first answered the call, I almost hung up. The man on the other end of the line would not tell me his name and wanted me to regard him as Deep Throat. I was busy and didn't want to deal with someone who had watched *All the President's Men* a few times too many. Just before I started to punch the "end" button on my phone, he told me that he had something from Pierre Salinger and that he had concrete proof that TWA Flight 800 was shot down.

The name Pierre Salinger kept my attention. I knew of him from his days as White House press secretary during the Kennedy and Johnson administrations. He later became known for his work with ABC news; he had covered some of the bigger stories of his time. He was known to be a maverick that was disliked by the anchors like Peter Jennings and Ted Koppel, but he had a good relationship with ABC President Roone Arledge. Salinger often used his position with Arledge to circumvent the news producers to make sure that the stories he was covering got the attention he wanted. But he stepped on so many toes at the network that he had to leave ABC in 1993. My Deep Throat explained that he had a video of a missile hitting a 747 and blasting it from the sky. "What do you need me for?" I asked. "Salinger knows what to do with a story like that; why in the world would he want to involve anyone else?"

It was too quiet; I thought the caller had hung up. Then his voice was back—calm, cool, and collected. "You don't understand the entire implication of what is going on here. Mr. Garrison, we'd feel much more comfortable to have you be the one who delivers this to the networks. I can assure you that what we have is real, and we really would like to see this get the exposure it should." I was still bewildered as to why he called me in the first place, but I took him up on his offer to view the tape and decide for myself whether it was worthy of taking to the networks. I got the tape a couple of days later, and it was everything the mystery man had promised. I could clearly see a missile hitting a 747 painted in TWA colors and the resulting fireball. Just like in the re-creations, the front of the plane fell right away as the aft section continued to fly before it broke up and fell from the twilight sky. I honestly believed that with all of the eyewitnesses confirming what one another had seen, the tape would be just the thing to get people asking the right questions to the right people to find out what had really happened the evening of July 17, 1996.

I know we all would like to have faith in the news industry and trust that the information they report to us is accurate, and that they make an effort to explore all possibilities to make sure what they report to us is correct. In the movies about news reporters, the mainstay of all the plots is how the hungry newsman will follow up on any and all leads to make sure the truth is told. TWA 800 was not a movie, it was very real. Two hundred thirty lost their lives. But what is also very real is that when I sent the tapes to the major news outlets, they conveniently lost the copy and asked that I not send them another. No one was interested in a theory that the plane was shot down, only that its fuel tank exploded. I still feel very safe when I fly on 747s. But I have to admit: I do worry about lights that rise from the ground in my plane's direction.

If the FBI did know of the Oklahoma City bombing a week before the explosion; if a missile, not a 757 crashed into the Pentagon; and if TWA 800 was shot down, the question that begs to be answered is *Why?* Why would the media organizations refuse to report on valuable pieces of these important stories? Did the news programmers not believe the evidence that was put in front of them? Did higher-ups in the government put pressure on the networks to not follow up on these leads? There are many questions, but few or no answers.

Speculation indicates the government had much to lose if these unreported facts came to light. Maybe some of the bigwigs were just covering their butts. Presidents Clinton and Bush would both have preferred to avoid public scrutiny of their administrations. If a Navy destroyer did indeed fire upon and cause the crash of a civilian airliner, resulting in the deaths of 230 men, women, and children, the Clinton administration would have been on the hot seat for years. And let's remember that Clinton was facing a tough re-election bid in a few short months. He couldn't afford anything to go wrong. Maybe it was just easier to blame an overheated fuel pump than have to explain a horrible accident. Maybe the FBI did know about the Oklahoma City bombing before it happened and just failed to take the right steps to prevent it. The country wants to feel safe and have faith in their government, and something like that would be very hard to explain to the public. As far as 9/11, it is more complicated. The implications that over three thousand people lost their lives, and the news that has and is still being delivered to the public may not be entirely true, is outrageous.

I don't know the full truth behind these stories. Answers, I believe, will be revealed to future generations. It wouldn't be the first time that facts of major events changed over time. For now, I know what I have seen and heard. If nothing else, there are parts to the stories that the press really should look into. It is their responsibility, with the trust that we put in them, to report what the truth is, even if we don't like the answers. I believe the people who have the power to decide what is reported on, at all the major news agencies, do take this trust very seriously, but I also think they have to answer to higher-ups in the political arena.

I have worked with Edward Lozzi since he left the press office in the White House. Ed is a public relations executive in Beverly Hills and handled publicity for Melvin Belli, as well as for many other high-profile people. I met Mel through Edward and became friends with him. He was not only the "Father of the American Trial Lawyers Association," but was also known as the "King Of Torts." Mel was privy to many conspiracy theories as legal counsel for Jack Ruby, tobacco litigation, and victims of secret U.S. radiation experiments, among other major stories. Mel taught me two things: "Don't believe everything you hear," and "To win in court, one must use visuals." I took his advice and sold my stories whenever possible with pictures, videos, and anything that presented a visual approach.

Mel told me a story once that was indicative of his genius methods. He had used Jack Ruby's dog to get him another trial. He explained to a jury that Jack left his best friend, his dog, in the car while going into the police station. Not only did Mel make sure that all the jurors were dog lovers, but he convinced them that Ruby would not leave his dog in a car with the intent to kill Lee Harvey Oswald. Mel won the trial, and his client subsequently died awaiting the new one. It was then that I realized how anyone could take a theory and twist it to prove a point.

Conspiracy-theory buff, I am not, but with what I have

seen and heard, along with the information my own investigations have uncovered, it is apparent that some of the news you see is really not news at all. Most of it is, and the media will always have the power of the trust you put into it. The news producers will always do their best to report the exact facts as they find them—perhaps as long as the higher-ups in the government say it's OK.

6

Ratings, Ratings, Ratings

Interview with Catherine Crier on the Robert Blake case.

Me with Christina Scheier during an interview on *Larry King*.

Christina Scheier being interviewed by CBS/*48 Hours* about Robert Blake.

Dave Holloway, father of Natalee Holloway, being interviewed by Larry King on the Aruba story.

The *New York Times* bestseller
I co-authored with Natalee's father,
Dave Holloway, and R. Stephanie Good.

Me with Larry King for a
Dave Holloway/Aruba interview.

At the Fox News studio for the Greta Van Susteren
interview on the Aruba case.

KIDNAPPED

LAST SEEN AT CARLOS & CHARLIES
MONDAY, MAY 29, 2005 1:30AM

NATALEE HOLLOWAY

CAUCASIAN AMERICAN FEMALE
BLUE EYES / LONG BLOND HAIR
5'4" 110 LBS. 18 YEARS OLD

ANY INFORMATION
PLEASE CALL 587-6222
OR CALL POLICE STATION 100

Natalee's missing persons poster.

High school pictures of Natalee Holloway.

Kent Walker and family, with brother Kenny Kimes, and mother Sante Kimes. I executive-produced the CBS movie starring Mary Tyler Moore. The book, *Son of a Grifter*, was an Edgar award recipient.

In the ABC News room covering the Pamela Rogers story. Rogers, a former teacher, was convicted of having a sexual relationship with a 13-year-old boy.

WANTED BY THE FBI

ARMED AND EXTREMELY DANGEROUS

Andrew Phillip Cunanan

DOB:	August 31, 1969
Sex:	Male
Height:	5' 10"
Weight:	160-180 pounds
Hair:	Dark Brown
Eyes:	Brown
Race:	White

Andrew Phillip Cunanan

THE CRIME

Andrew Phillip Cunanan has been charged with the murder of David Madson in Chisago County, Minnesota. Authorities believe that he fled Minnesota in Madson's Jeep Cherokee.

Several days later, authorities discovered the Jeep abandoned, on a street where Chicago, Illinois authorities were investigating a murder. Discovered missing from this crime scene was the victim's 1994 Lexus, which later turned up in a cemetery parking lot in Pennsville, New Jersey.

In Pennsville, New Jersey, authorities were called to a cemetery when the caretaker failed to return home from work on Friday, 5/9/97. The caretaker was discovered murdered. His vehicle was also reported to be missing. There are strong similarities between this murder and the other murders committed in Minnesota and Illinois.

The caretaker's vehicle is described as a red 1995 Chevy pick-up, with a dark interior, bearing New Jersey license plate KH993D.

REWARD: The FBI is offering $10,000 for information leading to the apprehension of Cunanan.

REMARKS

FBI poster of Andrew Cunanan, Gianni Versace's murderer.

DAILY VARIETY

FULL HEAD OF STEAM FOR

Proposed $20 Mil EEC Pot For Int'l Co-Productions Gets Euro Directors' Okay

PBS MARCH(ES) TO $36 MIL-PLUS PUBLIC PLEDGES

Sets Produc 116 Hrs. Of I Minis For Al

Oberstein Will Retire From CBS Records In June

NAACP Holds Hearing Into Charges Of Mismanagement At H'wood Branch

Scorsese Signed To Direct Beatty As 'Dick Tracy'

APTED SEEKS OF '28 UP' SOC

GARRISON MOVES FROM QUO STOX TO PRODUCING FOR TV,

GARRISON MOVES FROM QUOTING STOX TO PRODUCING FOR TV, FILM

Moving from a career as a Wall Street broker to a Hollywood producer, Larry Garrison is prepping several projects for the big and small screens, including "A Child Of Joy" with Dick Clark Prods.

"A Child Of Joy" is the story of March of Dimes poster girl Tracy Taylor, who was born with a host of birth defects, which she has overcome to have a relatively normal life.

Garrison became involved with the family several years ago and, impressed with her inspirational tale, optioned the rights to her story.

The Taylor story, perhaps starring the girl as herself, will be a telefilm for Clark Prods. in conjunction with Garrison Prods. Shannon Backley Garrison will be exec producer.

On the feature side, and the first project Garrison became involved

(Continued on Page 33, Column 4)

Front page article from *Daily Variety* headlining my company.

Me and Ellie Cook, Juror #5 in the Michael Jackson trial, getting ready for a television interview.

Taken with Anna Hauptmann and her lawyer, Robert Bryan. Anna was the widow of Richard Hauptmann who was charged in the kidnapping of the Lindbergh baby.

Melvin Belli and Jack Ruby during questioning. Belli was the attorney who defended Jack Ruby, the man who shot Lee Harvey Oswald.

Evidence and surveillance pictures
from the North Hollywood Shootout.

Diane Sawyer and Baby Jessica.

Shannon Buckley, Baby Jessica and family,
and me for an ABC *Primetime* interview.

An L.A.P.D. John Futrell card.
Futrell was the officer in the
North Hollywood Shootout.

Star Sightings:
Emmy Award winning Doris Roberts with my book, *Breaking into Acting For Dummies.*

Ben Affleck was photographed with a copy of my book and that picture appeared on the cover of *People Magazine* and in the *New York Post.*

Diane Sawyer and me on location.

See? I wasn't kidding. This really was the sign on the office door of Michael Schiavo's attorney.

R. Stephanie Good is the co-author of *Aruba* with Dave Holloway and me. She has been a news producer and consultant with me on many sweeps stories.

Chances are, no matter what your day consisted of, at the end of it you will tune in to the evening news in some form or another on your living room TV. The choice you make as to which program you watch will have an impact on how you spend the money you made today.

Comfort is perceived in having a familiar face tell you what happened in your world while you were out making a living. You might want to see what new news there is, or catch up on the stories you have been hearing about over the last few days or weeks. Chances are, you will also watch one of the news shows that goes into depth on current issues. You may or may not agree with the opinions of the guests on the shows, but the in-depth coverage, for the most part, is stimulating. Depending on what part of the country you live in, either right before or after dinner, there are a variety of choices to quench your news thirst. Viewers have a need to know.

Somewhere behind the talking head with the crystal clear voice is the news director. This person carries out a variety of different responsibilities. What is important to viewers is that the news is current, accurate and informative. The news director knows this, and he has at his fingertips a wealth of resources to make sure that he has the information viewers demand. It must be accurate, but also fast. Viewers do not want to wait for the latest developments.

Commercial breaks enable advertisers to tell you that you should buy their product or service, and they have thirty seconds to explain why. They use catchy little jingles; or have a famous, familiar face or voice quickly explain why the product they are endorsing is the right product for you.

Then, right before the newscast comes back on your TV, there will usually be a commercial to assure you that you have tuned into the right show to get your information, bragging about how well they do their jobs. Informative, accurate, and timely. It is the advertising spots that really drive what viewers see on television, and it is the Nielsen TV ratings that determine what they are worth.

Just about everyone has heard of the Nielsen Ratings. It is part of the information that Americans take for granted, and few have a real understanding of how the system works. Most of us place our blind trust in this system that determines what we and our families will have the opportunity to watch.

The Nielsen Company began in the infancy of the radio-broadcasting boom. Arthur C. Nielsen Sr. was astute enough to see that a system was needed to provide information to radio broadcasters and advertisers as to how many people were listening to any particular radio show, and who those people were. It was important information for the players of the time, as it was the only way to determine how much the radio show's advertising space was worth. Nielsen's company became the standard for how the radio industry determined if a show was a success or not and was already well-established by the time of the birth of broadcast television in 1950. It quickly became the official national measurement service of the television and advertising industries and has remained so for over fifty years.

The term *ratings* is misleading from the start. When the average person is asked to rate something, it usually means to offer a measure of quality, whether they like or dislike something on a scale of one to ten or how many stars, and so on. The Nielsen rating has absolutely nothing to do with rating the quality of a television program. The system does not care if the people who are watching a particular program are enjoying the broadcast or using it for background noise. The sole purpose for the company testing is to count how many people are watching, pure and simple. Then it goes back to the basic law of supply and demand. The more people who are watching at any given time, the more they can charge the advertisers for that time.

We demand fast, accurate information from our news; but success is measured by something else—a mere estimate of eyeballs. To add to the irony, every major news organization's goal is to capture the attention of close to 100 million households with televisions. About five thousand households, or thirteen thousand viewers, or one house out of twenty thousand, measure their success. Using so few to gauge the viewing habits of so many is hard to understand. How can the results be accurate? But it is the standard that advertisers and television stations have used for as long as the industry has been around, and there are no changes in sight.

Nielsen is aware that there are doubts about the accuracy of the system. On their web site they use this analogy to explain the process: If you want an accurate count of how many carrots, peas, and potatoes there are in a pot of stew, all you have to do is stir it up, take out a cup, and count the percentage of each vegetable. They say you don't have to eat the whole pot of stew to figure out what kind it is. According to Nielsen, it is not important to count all of the veggies in the pot to know how much of each is there. The logic that they use is that if you take enough small samples and count the amounts in smaller quantities, you get a good idea of the big picture. It sounds really simple, and it is a good metaphor.

Imagine that the American viewing audience is a big pot of stew. There are different types of viewers categorized by race, age, income level, and location—also known as demographics. Imagine that instead of a ladle dipping into the pot to scoop out a sample of vegetables, Nielsen scoops up little bits of information. The bits are gathered from electronic monitors hooked up to a few television sets, some of which have buttons for each member of the family to push when they are watching a show, or from personal diaries from a few people who write down what and when they watch. Nielsen then analyzes the small samples and calculates how many people from each group watched what television show. They break this information down into different categories: race, age, gender, income levels, and so on. With these figures in hand, they publish their findings and tell the rest of us what were the most-watched shows for the week and how many people from each category tuned in. These figures are critical to the people in charge of programming. A programmer may have the highest-quality, most critically acclaimed show in history on the line-up; but if it doesn't place well in the Nielsen Ratings the show will be a casualty, quickly.

Programmers on the entertainment side of the networks

have it easier than the news programmers. The entertainment side has weeks to prepare or adjust its schedule. The news programmers have to react at a moment's notice. On a hit sitcom or reality show, viewers tend to be loyal and continue to watch every week. In the news industry, things have changed. Viewers used to be pretty loyal to their favorite news telecast but today are more likely to surf the channels to get the information they crave. News directors know this, and they also know that their job security lies in the numbers the Nielsen Ratings provide. Far more people responsible for the news you watch have lost their jobs because of poor ratings than for any other reason. The pressure is constant, but is intensified during the time period that five thousand households supply the Nielsen Company with more raw data. Viewership becomes even more powerful. This is the time known as "sweeps."

During sweep periods my schedule goes from hectic to chaotic. News agencies are always looking for the next big story, but the sweeps are when the networks and advertisers pay the most attention. Two hundred and ten television markets in the United States are looked at under the proverbial microscope four times a year—November, February, May, and July. Every home that is part of the Nielsen group is sent diaries to keep a written record of what everyone in the household, including guests, watches for a period of one week. In some of the larger markets, sweeps take place three additional months—October, January, and March. The information that the Nielsen Company provides from these viewing times is used by decision-makers in the industry to decide what goes on the air and what is cancelled. It is also these times that determine what the networks will charge advertisers.

The stakes are high, because a small fluctuation in the numbers can add up to millions in advertising revenue. Needless to say, it is a very important time for the news agencies and not a very good time to have slow news days. If sweeps are coming up, my phone lets me know. It rings off the hook with producers who want the next "Oh, my God" story that will make the majority of the five thousand household diaries record the name of their show. There is greater urgency in the voices than in the off-sweeps times.

Sweep periods are my bread and butter. Agencies are hungrier for the stories, and I am in a much better position to negotiate on behalf of my clients. But it also contains some inherent risks. Breaking news events know no schedule. Whether it is a sweeps period or not, planes still crash and famous people still do naughty things. The majority of the stories I procure the rights to need to be aired quickly, if for no other reason than competition. When stories break during the crucial time, it's a no-brainer and one of my luckier days. The part that becomes a little complicated is when I get a story during off-sweeps times. Then I have to play the game that the networks hate. It is like cat and mouse.

After I have procured the rights to a story and am comfortable that no one else will run across the information, I can use it as a bargaining chip. I remember when I obtained the rights to the stories of two jurors who had served on the Michael Jackson child molestation case in 2005. For months the worldwide media had been transfixed with the pop star's trial. Entire news shows were dedicated to covering the daily developments of the trial, dissecting each event, from witness testimony to what people were wearing that day. You never know what a jury will do, but most of the news agencies were caught a little off guard when the jury found Jackson not guilty of all charges.

In mid-August 2005, one of my clients—Juror #5, Ellie Cook—stated publicly that she and another juror thought Jackson was indeed guilty and claimed that they were pressured to throw in their votes of not guilty. The story was far too big to hold back until the November sweeps, but I could still use the lure of the sweeps to my advantage.

Helping my client position herself for a book deal was my top priority, and I knew that the publicity generated would make it hard for publishers to resist. It was also important to Ellie on a spiritual level. She could have very easily stayed quiet and drifted off into the sunset, but she felt it was important that the truth be known. So I got my client on the air in mid-August and then again in November with the news agencies, knowing that I could easily produce more information when sweeps rolled around. We were actually the opening guests on the top-rated *Rita Cosby Live* show on MSNBC, and Rita's professionalism was well appreciated. My clients win: book deals are set into motion and they have a chance to make their statements. The networks win: they get the story and know that they will have even more for the sweeps period. With a story this big, it was a sure bet that when the sweeps rolled around, all Ellie would have to do is show up on camera to pull in better ratings—anything new she had to say would have been icing on the cake.

Some stories take on lives of their own. They capture the public's attention every time a new development surfaces. The networks love these piggyback stories to fill in the shows during the sweeps. But fillers, as they are known, also come in handy on the slow news days when the news agencies are hungry for anything to fill up airtime. By securing the news rights to parts of bigger stories that occurred outside of sweeps times, I can use this information to appeal to the news agencies for fillers when they need it most.

One of my more successful procurements was from the husband of an elementary school teacher who had a child with one of her students. When I first got involved with Mary Kay Letourneau's story, I really thought it would be a decent filler piece. I had no way of knowing it would become a story that would always be a sure bet to boost the ratings for the long haul.

Mary Kay and her husband, Steve, moved to Seattle in 1992 with their four children. He worked for an airline company and she was considered one of the most gifted teachers in the area. They were a well liked family who seemed to be living the American dream. That was until 1997, when it became public that the thirty-five-year-old Mary Kay was having an affair with one of her twelve-year-old students, Vili Fualaau. When it was learned that she was pregnant with the child of the early teen, the press went on a feeding frenzy.

I procured the news rights of her soon-to-be ex-husband and had firsthand knowledge of how devastating the situation was to her family. The press paid attention. Sad stories add to the ratings. Showing the anguish of a new, single father with four kids to take care of alone because his wife decided to become romantic with a twelve-year-old, well let's just say it was a story the sweeps couldn't resist. When the coverage focused on Mary Kay, it became a free-for-all.

The media became transfixed by the day-to-day events of

the story. If it had been a male teacher who had impregnated a twelve-year-old girl, the news coverage would have been very different. But, it was a woman who had abused a teenaged boy, and the media had something new to report on—a female pedophile. This story stuck around for years. It was one of the stories that could be used in the sweeps because it had a proven track record. It also came in handy on the slow news days to keep the viewers tuned in.

Mary Kay was charged with second-degree rape of a child, to which she pleaded not guilty in March of 1997. She was released on bail on the condition that she would not have contact with her four children without controlled supervision, and under no circumstances was she to have any contact with her then thirteen-year-old lover. Two months later she gave birth to the child that Vili had fathered, and the baby was eventually taken away. All the evening network shows covered the story. My phone rang off the hook with the voices at the end of the line buzzing, "Hey, Larry, got anything else on the teacher?" The news programmers were on pins and needles waiting for the trial.

In August 1997, the now ex-teacher changed her plea to guilty. She would go on record citing her full responsibility for her actions, all along saying she wanted to spare her young lover from the ordeal of having to testify. No one knew the name of the thirteen-year-old until *The Globe* revealed it to the country years later, for a fee that was rumored to be about fifty thousand dollars.

Developments in the sentencing phase of the case were enough to keep news agencies' attention even though they lost out on the chance to spin the testimony in a full-fledged trial.

Mary Kay's attorney, David Gehrke, presented a shopping list of mental illnesses that his client suffered from; and, to throw in a dash of drama, the fact that her own father had fathered two children out of wedlock with one of his students years before. Commentators had a smorgasbord of subject matter to collaborate on, and the news cameras focused on a type of face that you don't often see in the hot seat before a judge. Mary Kay was not a celebrity, serial killer, or a hardened criminal. She was attractive and had a pleasant, honest demeanor that was hard not to like. Even her statement to the judge came across as admirable, in a pathetic way. "I did something that I had no right to do, morally or legally. It was wrong and I am sorry. I give you my word it will not happen again. Please, please, please help me. Please."

With her estranged husband at her side, who had since relocated to Alaska with their kids, and the victim's mother also in the courtroom saying that her son's lover's conviction would serve only to make her son feel guilty, Mary Kay Letourneau really did look like someone who needed help. The tabloid shows had a field day. Shows like *Inside Edition, Hard Copy,* and the like had hit the ratings payload. People who knew Mary Kay when she was younger and parents from the school where she taught all received a little limelight as the American public craved more information about this strange case.

Mary Kay's attorney earned his retainer. She was sentenced to seven-and-a-half years initially, but the judge was convinced to reduce the term to six months, as long as Letourneau underwent treatment for sex offenders, medical attention for bipolar disorder, and agreed to have no further contact with Vili. For all practical purposes, the story seemed destined to die with the rest of the tabloid stories that capture the ratings. But this was no ordinary story.

Mary Kay probably would never have been heard from again, were it not for her actions after her release from jail. About a month after her release, a Seattle cop stopped and checked out a car that had its lights on in the rain with very steamy windows. Inside he found Mary Kay with Vili, about six grand in cash, and her passport. After the arrest and investigation, it was learned that the couple had indeed resumed their relationship right after her early release from prison. News programmers pray for stories like this; they are just too easy to pass up. The couple found themselves in the public eye once again, and this time the judge was not as understanding.

After hearing testimony of how Letourneau had ignored her probation orders never to have contact with Vili again, and had made excursions to Pike Place Market in Seattle and gone to a public theater to watch the movie *Titanic,* the judge reinstated her original sentence of seven-and-a-half years. What caught the press' attention is that she became pregnant with the teenager's second child during her brief break from jail.

No one could have predicted how much airplay the teacher from Seattle, who had a torrid affair with her twelve-year-old student, would generate—or for how long. Twice the story seemed to die. First, when Mary Kay went to jail for the first time and second when the media reported that she gave birth to Vili's second child in prison. The story seemed to be over, as far as the media was concerned. But it wasn't the last we would hear from Mary Kay Letourneau. There were a few "Where are they now?" news pieces run on the smaller news shows over the next seven years. There was a Court TV interview with Mary Kay's cellmate, but for the most part, the ratings wars were looking elsewhere to drive up the value of their advertising space.

On August 4, 2004, Mary Kay Letourneau was released from jail. By this point, her victim had come of legal age. The next day, Vili requested that the judge lift the order preventing contact between the two. The judge complied, and round three of the Vili/Mary Kay/Nielsen's ratings romance was back in full swing. The news talk shows had new ammunition for an old gun to fire into the American viewing public, complete with old video and new controversy. In mid-September 2004, it was rumored that the two were engaged. The official announcement of wedding plans took place on October 12, 2004. It looked like the November sweeps had their story.

There are times when the attention a story receives from the press actually becomes part of the story itself. A story that had outraged the country seven years before became a curiosity story that engulfed the nation. Mary Kay's picture was plastered on magazine covers and the news shows were scrambling for interviews from anyone even remotely related to the couple. There was even an on-line registry that the public could check out. A few people even bought gifts.

The couple wed on May 20, 2005, at a winery in the Seattle suburb of Woodinville, in a ceremony under very tight security. The bride was in her early forties and the groom was twentyish. Their courtship had lasted over seven years. At the event funded by tabloid-type media outlets to capture the final chapter of this story—at least I think it's the final chapter—a large part of the budget was spent to make sure that uninvited media and paparazzi were kept out.

Sensationalism in one form or another has become a standard in television news. That fact troubles many of the news programmers who take the job of delivering information to the public very seriously. Even if a news programmer wants to focus on events that a vast majority would concede are important, they can't. My friend, who is a programmer of a major news network, said that if he does not use the stories that are really tabloid in nature, like Mary Kay Letourneau's, his competition will, and he will lose his audience to them. The entity that is responsible for keeping the public informed as to what is happening in their world is at the mercy of those five thousand households who control what the rest of us watch, and they seem to be fond of tabloid stories.

7

That's Entertainment

What makes a news piece an "Oh, my God" story? Something you hear about and can't help wanting more details. There are no black-and-white answers; it's more of a feeling that it provokes in viewers. It might start out as idle curiosity and fester into a burning desire to know more. It has to be out of the ordinary and fairly shocking. Although it doesn't have to be, it usually has to do with a crime or tragedy, and involves aspects that can be hard to believe and defy understanding.

"I should write a book!" has been said by many, and most of us have been acquainted with someone who has had an interesting life whose memoirs would make a good read. The problem is getting it published. The same applies to the film industry, whether it is for television or the big screen. Interesting is not enough—it has to be shocking and visceral. For every thousand good stories that are presented to the networks, studios, and publishers, one will make it to production or print.

Twenty-five years in the industry has provided me with my fair share of the "Oh, my God" stories. But, to illustrate the process of having a news piece evolve from being aired on the TV news shows to growing into a book and film, it is easy for me to pick one out. Just the words that offer a brief description of the subject matter grab your attention: arson, fraud, murder, millionaires, slavery. When you add the fact that a mother and son are being described, let's just say it puts the "Oh!" in "Oh, my God!"

Wires were buzzing in July 1998, with a story about a mother-and-son con team that had been arrested in New York City. They had originally been picked up for bouncing a check at a used-car dealership in Utah, but within days, they became the tabloid headlines in the Big Apple and the news scoop of the nation. It is a complicated story.

When I was searching the wires about the story, I found that the two had been linked to the disappearance of a wealthy socialite by the name of Irene Silverman. Silverman was an eighty-two-year-old widow who owned a townhouse that sat on some of the most valuable real estate in the country. It was just a couple of blocks from Central Park in Manhattan's Upper East Side. Irene Silverman was known as a little eccentric and colorful woman who had converted her townhouse into apartments for the rich and famous.

In the month prior to the arrest, a twenty-three-year-old man by the name of Manny Guerin showed up on her doorstep to rent one of her posh apartments with fake references and six thousand dollars cash in hand. Guerin's assistant, a woman, had called Silverman earlier to announce his arrival. Normally Silverman would have checked out the references, but the lure of the wad of cash got the best of her, and she quickly had a new tenant. It would turn out to be the mistake that would cost her her life.

Irene Silverman went missing on July 5, 1998, but that was not what grabbed the headlines. That was still a few days away. If a wealthy old woman disappears, it may get a quick mention on the local newscast and maybe a spot like "If you have seen this woman or have any information on her whereabouts . . ." It was noticed that her newest tenant was also missing. There was no evidence of foul play; it was as if they both had just vanished.

A few blocks away, on the same day, a task force that included the LAPD, NYPD, and FBI placed a mother and son under arrest. The official charge at the time was the purchase of a used Lincoln Town Car with a bad check. The two stories seemed to be completely unrelated, until an officer from the task force saw a police sketch of Manny Guerin on the local evening news. He picked up the phone and called the NYPD to let them know that Manny had been in custody for a few days, along with his mother; and his name was not Manny Guerin, it was Kenneth Kimes.

When the two stories came together, the press took notice and started looking into the background of Kenneth and his mother, Sante. Once reporters and news programmers had a glimpse of the mother's history, the story was destined to be the "Oh, my God" news piece of the year. Within days, the con team dominated the headlines. The *New York Times* did a huge front-page story on the pair, as did many major papers across the country. Every major television news outlet was scrambling to be the first to broadcast the story of the mother-and-son con team. Within days, they were the subjects of segments of both ABC's *20/20* and NBC's *Dateline*. For shows of this nature, it would normally take a few weeks to put together a segment, but the media monster couldn't wait to get the word out. The story was just too bizarre, and the competition for it was fierce.

Sante Kimes was reported as being a con artist on a large

scale for over thirty years. Her criminal career started off fairly small in the early 1960s, with a series of shoplifting charges and credit-card fraud cases. She had been arrested on many occasions but never served any real time for her misdeeds. People who knew her from that time period described her as a beautiful Liz Taylor look-a-like with a charismatic personality, and a very dark side.

She then married Kenneth Kimes, a successful semi-retired developer, whose net worth was estimated to be $20 million. Most observers would have thought that Sante's crime spree would be over, being married to a multimillionaire, but the crimes only escalated. And now her husband got in on the fun.

In the mid 1970s, during the celebration of the nation's bicentennial, Sante conned herself into Washington D.C. political parties by claiming that her husband was the honorary bicentennial ambassador. The con worked so well that she was able to swindle her way past the secret service and have a picture taken with Vice President Gerald Ford and his wife. The agents assigned to protect the vice president were caught off guard but quickly escorted the Kimeses out of the building after the picture was taken.

That didn't stop the Kimeses. They went on to crash several more parties in the D.C. area that night, including a couple held by real ambassadors. Their actions did not go unnoticed; the pair ended up in the *Washington Post* the next day with the picture of them and the vice president. The article described how Sante had conned her way into the White House and had her picture taken with the first lady, Pat Nixon.

Not long after, Sante Kimes and her husband were arrested for stealing a twelve-thousand-dollar mink coat. They didn't take it from a store; they were having drinks at an upscale lounge in one of Washington D.C.'s high-end hotels. When Sante got up to leave, she put her coat over another woman's, scooped them both up, and left the room. Another patron of the bar witnessed this and reported it to management, who in turn reported it to police. The Kimeses were handcuffed in their robes in their room at the Mayflower hotel and hauled off to a night in jail.

Multimillionaires who crash political parties at the White House then go out and steal fur coats would have made an incredible story, but they were just getting started.

Sante was convicted on slavery charges in the mid 1980s, the only case of its kind in the twentieth century. She had enslaved Mexican girls that were in the country illegally to do domestic chores like cleaning, cooking, and ironing her clothing. She served four years in federal prison for the crime, at the same time fighting criminal charges of the fur theft and civil charges from the maids she had enslaved.

News shows paraded several people who had been associated with the Kimeses over the years and attorneys who had dealt with them. One of the attorneys said that Sante Kimes had gotten her insurance companies to fork over a million bucks to settle the civil cases the maids had brought against her. It wasn't that the insurance policies covered slave holding; it was that she had just worn them out and intimidated them by threatening some of the bigwigs' family members. One described how he had found a dead crow by his car the day after Sante Kimes showed up at his home, threatening to harm his son if the insurers didn't pay.

News reports also described how the Kimeses' homes had

a habit of burning down. In the late seventies, Sante's beach-front home in Hawaii had burned from an arson fire, then the same home burned again in 1990. The first fire provided the millionaires insurance money to add a pool and substantially upgrade the house. The insurance companies didn't pay the claim on the second fire because of the suspicious nature of the fire and the first of Sante's possible murders.

Lawsuits were a staple of life for the Kimes family because of Sante's actions. They hired some of the best criminal and civil lawyers in the country to defend them, but they also used attorneys that were down and out. This provided Sante the opportunity to play lawyer herself, as she coached the lawyers whose only office was the kitchen table in their one-bedroom apartments. Her only real use for them was their licenses to practice law. One of these small-time lawyers was a man named Elmer Holmgren. He disappeared without a trace the day after he told ATF agents that the Kimeses had paid him to burn down their house in Hawaii the second time.

Kenneth Kimes Sr. died in 1994. Most of his fortune had been eroded away by endless lawsuits and a lifestyle that included homes in Las Vegas, Hawaii, and the Bahamas. Sante Kimes found herself in a financial crunch and, worst of all, there was no will. Whatever was left of the Kimeses' fortune was out of her reach.

Reports went on to tell how Kenneth Kimes Jr. then became her partner in crime. Kenny, to those who knew him, was in college at the time of his father's death. His relationship with his mother was reported to be tremulous during his adolescence. Friends of the family who were interviewed described Sante as a controlling and demanding mother. Kenny was not allowed to attend public schools until his mother was serving time for the slavery charges. Until then, private tutors home-schooled him eight hours a day. She interfered with his social life, telling people that Kenny wanted to be friends but that they were not good enough to play with the millionaire's son. Every interview that described the relationship between Sante and her son described one where Kenny had learned to hate his mother over time. Everyone assumed that once he went away to college he would be free of her controlling tendencies and that he would have a shot at a normal life. But something inexplicable happened after Kenneth Sr.'s death. Kenny dropped out of college and became his mother's partner in crime. This is where Sante's already substantial criminal record expanded to include murder.

Mother and son traveled to the Bahamas in search of what was left of Kenneth Sr.'s fortune. Before his death, Kenneth Sr. had placed money in offshore accounts to protect it from the endless judgments as the result of the barrage of lawsuits he had to face because of Sante's actions. The problem was, the bank he chose to hold his money was going under, and Sante was trying to gain control of it by using a variety of forged and illegally notarized documents. Syed Bilal Ahmed was an executive of the Gulf Union Bank in the Bahamas, where Kenneth Sr. had put a substantial part of his remaining fortune. Sante and Kenny had dinner with the bank executive on September 4, 1996, and it is believed the purpose of the meeting was to find a way to gain access to the money. That was the last time that Ahmed was seen or heard from.

Bahamian authorities launched an investigation into the banker's disappearance. When they learned that the last people to see Ahmed were Kenny and Sante, they made attempts to interrogate the pair. It was too late. The mother and son had left the country and fled to California before the Bahamian authorities could question them.

The crime team was scrambling for ways to get cash and came up with the idea of getting a mortgage on the family home in Las Vegas. The problem was that they got the mortgage in the name of a family friend without him knowing about it. David Kazdin had been a friend of the Kimes family for years. He had allowed the Kimeses to put their Las Vegas home in his name in the 1980s, making him the legal owner of the house, in an attempt to protect the property from seizure. They would transfer the property, using a quick deed of trust, making it look like Kazdin was the legal owner; then when the heat was off, they transferred the title back into their names. It was a game they played for years.

David Kazdin learned in 1994 that he not only still owned the property, he had also taken out a loan on it in the amount of $280,000.00. Sante and Kenny Kimes had forged documents that put the money in their pockets, and Kazdin was left responsible for paying the debt. If that wasn't bad enough, he later learned that ownership had been transferred once again into someone else's name, then the house was burned by an arson fire nine days after an insurance policy had been purchased. Kazdin told family and friends about what had happened just before his body was found in a trash bin outside of the Los Angeles International Airport. He was killed by a single gunshot to the head, execution style.

The LAPD uncovered the fire/mortgage scam in their investigation into Kazdin's murder and were on the hunt for

Kenny and Sante. In their investigation, they happened across a man named Stan Peterson, who had been involved with the Kimeses and had sold guns to them illegally. Peterson called the LAPD and told them that he had received a call from Sante Kimes. She instructed him to fly out to New York so he could run an apartment house in Manhattan that she was buying from a wealthy widow who wanted to travel the world. The authorities set up a sting and had Peterson go to New York to meet with Kimes.

Agents from the LAPD and FBI kept Peterson under surveillance, and when Sante and Kenny approached him he took off his ball cap, the signal law enforcement and Petersen had agreed upon beforehand, and the mother-and-son team were taken into custody.

Arresting the Kimeses on July 5, 1998, proved significant, as that was the same day that Irene Silverman disappeared along with her new tenant. But law enforcement failed to make the link, even though Sante Kimes had Silverman's passport and credit cards in her purse at the time of her arrest. Sante and Kenny had plenty of problems already with the law, but they must have felt that they had gotten away with the Silverman crime—at least for a few days. As I mentioned before, it was a matter of luck that an officer from the task force that had arrested them happened to see the local news and make the connection that Kenny and Manny were the same person.

A mother-and-son con team: The mother was known by over fifty different aliases and lived the life of a millionaire—while stealing fur coats and enslaving people in her home. A twenty-three-year-old man, linked to three different premeditated murders along with his mother, with a history that included multiple arsons and cons that reached all the way to the White House. I have been in this business for a long time, but I had never seen a news piece with so many twists and turns and loaded with so many sensational hard-to-believe facts. Stories like this don't come along often, and I knew that the news media would pay attention to the Kimeses for a very long time. I also saw the potential for a book and movie deal.

I combed through every newswire and article I could get my hands on to learn more about the pair and to find a way to be involved in the development of a movie on their story. I remember a couple of newspaper reports that called Sante "The Queen of Cons," but the *New York Post* gave the mother-and-son con team the tag that would stick: "The Grifters." In the early '90s, John Cusack and Angelica Houston starred in a movie with that title. It was about a mother and son who were con artists, and it was impossible to ignore the similarities in the film to the real-life Grifters.

When a story gets the type of attention that the Grifters were receiving, and I want to find a way to lock up at least part of the news rights, I have to find something my competitors have overlooked. Maybe a friend or someone who went to school with a subject of a story can be found, and I can get some inside information that has not yet been reported. It had worked before, but this story was receiving so much attention that I had my doubts that anything new could be discovered. Then I stumbled upon an article in the *New York Post* a few days after the media blitz had started. They dedicated a section in their paper to the Grifters that included interviews with Kenny's tutors from years before and some new revelations about Sante. What caught my eye was a small article about another son. Sante had an older son from a previous marriage who was doing everything he could to stay out of the public eye. The article described a family-oriented man who sold vacuums for a living in Las Vegas. It even had a picture that was taken outside of his office without his knowledge. It was clear the article was a product of reporters talking to people who knew him and not by interviewing him. The only quote from the other son was, "I want nothing to do with this."

Sometimes more than just my professional curiosity gets the best of me: besides just wanting to get the rights, I wanted to know more for myself. After spending days pouring over every bit of information I could find about the mother-and-son grifters, the fact that there was another son who was selling vacuum cleaners while his mother and brother were stealing millions—well, that just made the story all the more captivating. I decided to try and talk to the "other son" and see if he would be interested in having me represent him and maybe shed some light on the rest of the story that the public may not yet have known.

The article gave me his name—Kent Walker—his location, and his type of work. That was all I needed to track him down. I found a Las Vegas Yellow Pages and started to dial all of the vacuum stores in the city. After about half a dozen tries, I hit pay dirt. A secretary answered the phone, and when I asked for Kent Walker, I could tell that she was uncomfortable. It was apparent she had been fielding a lot of calls from strangers wanting to find out more about the Kimes family. She put me on hold and in a few moments the other son of Sante Kimes was on the line. There were no hellos or formalities, just a straight-to-the-point statement. "If you are a reporter, the only thing I have to say is no comment and please leave me alone." Not exactly the conversation I was hoping for.

I started to explain that I was not a reporter and that, in fact, I wanted to help him deal with the media. Then I made a mistake. I was trying to figure out a way to gain his trust and knew that everyone else trying to contact him was most likely assuming that he was a criminal like the rest of his family. He would want to stick up for them, so I said, "Look, I know that the press can exaggerate things. I would like to give you the opportunity to have the truth told about your mother and brother." With that he started to laugh.

"Mr. Garrison, I just want this mess to go away so I can get on with my life," he said. "I couldn't really care less if the truth about my family ever comes out. In fact, I'd just assume it didn't. I want to be as polite as I can right now, but I have nothing to say to you or anyone when it comes to the subject of my mother and brother. I'd rather be left alone."

I could tell by his voice that he was firm in his conviction not to talk to anyone, and I didn't have the heart to tell him that the situation wasn't going to just go away. Right before he hung up, I gave him some of my credentials and urged him to write down my name and number. Then I offered to help him in any way I could when he was approached by the media or studios wanting his story. I explained to him that he would be getting offers for his film and literary rights, and whether he went with me or not, I would be more than happy to help him check them out to be sure that he was getting a fair deal. I don't mean to sound naive, but I could tell in the first few moments of the call that Kent was sincere. He was not interested in trying to gain from his family's newfound notoriety; he just wanted to get his life back to normal. As we hung up, I was not sure if he had written down my contact information or not. In any other case I would have pushed a little harder to attain his news rights, but I backed off. I didn't want to push him.

Even though I couldn't get Kent Walker to work with me on the Grifters story, I did get involved with the lawyers who were representing Sante and Kenny. In most stories like this, and especially when the subjects are behind bars, it is a given that you have to work through the attorneys if any kind of deal is to be made. The defense team that represented the Kimeses had no problem with me helping them put together a book and movie deal, as long as the book and movie would be about them and how they represented the mother and son. The lawyers wanted to take advantage of the media the story was generating to further their own notoriety and legal careers. It has happened many times before. When you represent high-profile clients, your legal practice benefits from the public exposure. Many times, the lawyers themselves love the attention of the limelight.

The Kimeses wanted to fight their case in the eyes of the public and stressed the injustice that was being done to them. I started working with the team and even spoke with Sante and Kenny a few times, while trying to position their story in the press. Our conversations added up to hours of details on how they were framed and were the victims of a corrupt system that was out to get them. Although I still saw the commercial value of their story, I knew in my gut that the Kimeses were not being straight with me; and their attorneys were in it only for the publicity, not to truly defend their clients.

Several months passed and I was making some progress in getting them the chance to get their story out, when I received an unexpected phone call in February of 1999. "This is Kent Walker, do you remember me?" I almost dropped the phone when I recognized the voice of Sante's older son. He told me that the press had eased off of him over the months, but the damage was already done to his life. The Grifters were still receiving a lot of attention from the media, and when people realized that he was also Sante's son, they started to treat him differently. He explained some of the challenges he had in trying to protect his kids from the media circus, and that he felt he was failing. The situation had also affected his ability to run his vacuum business efficiently, and it was starting to take a toll on him financially. One of the attorneys who represented his family in the slavery cases was well-known in Hollywood circles and had tried to put together a possible book deal, to no avail.

Right away I could tell a difference in his voice from the first time we had talked. On our first contact he came across as strong and decisive, now he just sounded tired. I explained to him that I would be happy to represent him and told him that I had been in negotiations with his mother and brother's lawyers to represent them also. That's when he dropped the bomb on me. "There is no way you can represent the three of us. If you're going to help me with this you have to remember one thing: I am the good guy. I will have nothing to do with being represented by anyone who has anything to do with representing them."

From the outside, it might have looked as if it was in my best professional interest to continue to represent the Kimes team. After my conversation with Kent Walker, however, it became clear that he had the story that would capture the truth, and hopefully commercial success. I agreed to devote my efforts to the story from his point of view. Before we hung up, I had one question for him that was nagging me. "What made you decide to call me?" I asked. He told me that he had decided to put everything behind him and try to just go on. In his desk he had kept the cards and numbers of everyone who tried to interview or represent him. As he was throwing the contact numbers away, he came across mine. "I remembered you were the only one who didn't pressure me" he answered. "That's why I called you."

I arranged an interview with *Extra*. I had done numerous sweep pieces with the show and knew they would be interested in the story. The purpose of the meeting was to see exactly what Kent had to say about his family and if he really did have any new information that had not been already reported.

Kent arrived on a flight from Las Vegas to Burbank and I met him at the gate as he got off the plane. I remember feeling a little nervous at first. He was a large man with long black hair, a full beard, and a deep voice; he seemed a little rougher than I had imagined. I was encouraged that he had a look that would work well on video, but still wondered if he was more like his mother than he claimed to be. As we drove from the airport to the studios of *Extra,* I explained the process to him: After the initial interview, the producers would let me know if they were interested in doing a piece on him, then I would get back to him to schedule. As we walked into the business offices of *Extra,* I could tell that Kent was nervous, but he held his composure as we were seated in a room with several of the news producers. Then he started to tell his story. An hour and a half later, I knew I had picked the right person to represent the Kimes story.

In a room full of strangers, Kent went into detail about

what it was like to live in the shadow of Sante Kimes. He explained his thoughts on why his brother chose to live a life of crime with his mother instead of going out and living honestly, as he had. He also revealed the guilt he carried for not having tried hard enough to pull his younger brother from his mother's grip. Kent was able to illustrate how Sante was able to get away with so much for so long, just from the sheer force of her personality and her uncanny ability to outsmart the legal system. He also went into detail about the dynamics of the relationship between Kenny, his mother, and his stepfather, and also shed some light on some of the reasons why Kenny had turned into a murderer. The situations were bittersweet. In one breath he would describe moments of violence that were almost beyond belief and in the next tell of fun-loving times he and his family had shared. Kent also filled in some of the pieces of the story that had been reported by the media already and came across as an honest, sincere man when he chronicled the life and times of the now infamous Kimes family.

As the initial interview started to wind down, I knew that his story had what it took to become a great book and saw the potential for a movie. What surprised me was when I looked around the room. We were surrounded with people who cover stories like this, or close to this, for a living. People become stoic over time when covering crime pieces and learn how to keep their feelings out of their jobs. I saw a couple of the interviewers drying their eyes as Kent finished up.

Immediately the producers approached me and told me that they did not want to wait to do the interview; they wanted it right away. I convinced Kent to stay the night, and the interview was conducted the next day at the Universal Sheraton in

Universal City, California. Although I was pleased with the results of the interview, my job was just beginning.

Challenges were inevitable. First of all, the Kimeses were still in the middle of an active murder investigation leading to trial. Publishers and studios sometimes are leery to do any story until the final outcome of the criminal process is known. Another big concern was public domain. The story of the Grifters had received an enormous amount of coverage. They even made an appearance on *60 Minutes* after I had cut off contact with them.

Anyone can write a book about the subject, once stories like this become public, and it happened with this story—twice. The first book was published not long after I started representing Kent. It seemed thrown together from different news reports and offered a surface description of the events that led up to the Silverman disappearance. It was also announced that another book was being written by one of the reporters who was covering the story. ABC was considering doing a TV movie based on the Grifters, so it was a race to see if I could put together a film deal in time.

Kent was interviewed on *Good Morning America* by Diane Sawyer as I worked to put the deals together. The interview was partly to help get the word out for Kent that he was not in the same category as his mother and brother, and partly to get him the exposure I needed to pitch a film and book on his life.

Months passed, but after a lot of persistence I found a book agent in New York who saw the potential in Kent's memoirs being published. She hooked us up with a writer to help him tell his story. When he found out that he would have to share the by-line with Kent, and that it had to be written in first-person from Kent's point of view, the writer backed out. He was more interested in writing about the escapades of Sante and not in telling the story behind the headlines. It looked as if we were done with any chance of a book deal, but I didn't give up. A couple of months later, I found another agent who had an award-winning writer willing to let Kent tell his side of the story, instead of repeating what the tabloids had already reported. After months of work in the form of interviews and writing up a proposal, we got our book deal. HarperCollins saw the commercial potential of the project and published *Son of a Grifter* in April 2001. The book was both a commercial and critical success. It reached number 17 on the *New York Times* Best-Sellers list and received great reviews from around the country. What I am most proud of is that the book was the recipient of the 2002 Edgar Allen Poe Award for Best Fact Crime.

At the same time that Kent was writing his memoirs, I was working on getting his story told in film version. Like the book, it seemed that the odds were against us being able to develop a TV movie on Kent's story. Other networks already had plans in the works to do a movie on the Grifters, and the criminal charges the Kimeses faced looked like they would drag on in court for years to come. I approached other networks with the proposal of doing a movie, but they wanted to steer clear. Then ABC announced that they were backing out of doing their movie on the story. The chances seemed a little better, but the challenges were still many.

Our break came when we learned that Mary Tyler Moore was interested in playing Sante Kimes in a TV movie. That got CBS's attention, and they went forward in developing *Like*

Mother, Like Son. My partner of twenty-five years at Silvercreek Entertainment, Scott Brazil, and I were pleased with the film we executive produced. Although the final cut of the movie was more of a story of the Irene Silverman murder (Silverman was played by Gene Stapleton) and the relationship between Kenny and Sante, Kent was still a consultant for the film. He was able to provide information that ensured the theme of the film was based on fact instead of the tabloid headlines. The movie turned out to be one of the highest rated TV movies of the year for CBS. It beat out *The Diary of Anne Frank* and *The Sound of Music,* and is still played on the Lifetime channel to this day.

Sante and Kenny Kimes will spend the rest of their lives in jail. Even though Irene Silverman's body was never found and there was no direct evidence linking the two to her disappearance, they were both convicted of her murder and sentenced to over a hundred years in prison, each. They were then extradited to California to face charges in David Kazdin's murder, where they both faced a possible death penalty. Six months before the trial started, Kenneth Kimes agreed to come clean in return for the death penalty being thrown out for himself and his mother. During Sante's trial, Kenny confessed to the murders of Irene Silverman, Syed Bilal Ahmed, and David Kazdin, and implicated his mother in all three murders. Sante was returned to New York to live out her life in prison, and Kenny will serve his life sentence without the possibility of parole in California.

Of course, not all stories that become books or subjects of movies are as dramatic and complicated as this story. I chose to use this piece to illustrate the process of a news story becoming a book or movie for a few reasons. First, it is one of the highlights of my career; but it also illustrates how much luck plays into the final results. If Kent had not called me back, the book might never have happened, and the movie would have been much different. When I first started to represent Kent, I had to talk him into doing a book. He didn't see the potential; and if he hadn't stumbled on my number when he was cleaning out his desk, the book would have never happened.

We were also lucky in that a big-name star like Mary Tyler Moore was part of the movie. Had it not been for her interest, I am not sure the film would have made it out of the development stage. There are so many variables when it comes to a news story becoming successful in the entertainment industry, and having a good story really isn't enough.

It also illustrates the risk of movies and books that use only information in the public domain as their source. So much of what was reported about the Kimeses was inaccurate, including allegations of incest. The media have to use spin to stay competitive. I feel it is important that publishers and studios use everything that is available to them to ensure that the stories they tell are accurate. Having someone who was part of the story helping to tell it adds to the credibility and overall honesty. I have had clients who had their stories told from public domain without their input; for the most part, the stories are sensationalized beyond the facts of the case.

Personal interest is the main reason I chose to use this story. I have become friends with some of the people I have represented over the years, but Kent and I have enjoyed a close friendship since the book hype was over and the movie has played out in reruns. I found his story to be inspirational. He is a true survivor, and he showed me that you can do the right thing no matter what your environment. I am also proud of the fact that I was able to help him tell his story so it could help other people who may have been trapped in similar situations.

Sharing personal losses at the same time also created a bond between Kent and me. When he lost his brother to the prison system, I could feel the pain it was causing him. He did not want to see Kenny go free, but he still felt the loss of his baby brother. While this was happening in his life, my brother passed away unexpectedly. We helped each other through rough times in each of our lives. We also found ways to keep our senses of humor. I purchased a home with a beautiful view of a lake—the only problem was, the property across the street had two very large trees that blocked part of my view. I remember asking the son of a grifter, in jest, if he could arrange to have them removed, to which he replied, "I'll have the boys get right on it." The trees are still there.

In this business of news, and the books and films that come from it, it is sometimes hard to find the truth. It can be even harder to find someone you can trust. In Kent Walker, I found both. He had the courage to come forward and tell the truth about his family without trying to make himself look like a hero. I also found a friend I can trust, and he inspired me to write this book.

8

Caught in the Act

My adrenaline was high and my heart was racing as I navigated through the streets of Hollywood. I was running late for a major studio meeting. Glancing up, I saw the signal light at the intersection ahead had just turned yellow. The car in front of me had to make a split-second decision: play it safe, stop, and be a little late getting to where you need to be or take the chance and hit the gas to try to beat the red light. That driver put his foot to the pedal and a flash was evident. In the old days, the odds were on your side that there wouldn't be a traffic cop nearby, just in case the light changed to red before you cleared the intersection. Now you have to think about Big Brother. Living in Southern California, I have experienced the scenario countless times. Every time I see those cameras mounted in busy intersections I have to smile a little and say to myself, "What have I done?"

Surveillance cameras are at intersections to see if you run red lights, in stores to prevent shoplifting, airports for security, just about every public place you can think of for a variety of reasons. I am not entirely comfortable knowing that there is always someone looking over my shoulder. Having said that, I also have to admit that I feel a little responsible that they are there in the first place. In the early 1990s, the use of these cameras was still fairly novel, as the technology was too expensive for the average person or business to afford; but they were still around. I found a way to use the video shot by these undercover mechanical spies to entertain people. It was an idea ahead of its time.

At that time, I partnered with Dick Clark, his film group, and production company. My office was only steps away from his, and I still remembered the excitement I felt growing up in Long Beach, New York, waiting for *American Bandstand* to come on and counting the days until Dick ushered in the new year. My family huddled around the TV set in our living room as Dick introduced the acts, and nothing could have pulled us away.

When I stepped into his office in the early 1990s, I was surrounded by the memorabilia from his years in the music business. Mementos from Elvis, the Beatles, and countless other acts were in every part of the building. Working with Dick was surreal to me. He and his wife were beautiful, down-to-earth people, and after I got over the feeling of being a little star-struck, I really enjoyed my time working there. I had a blast. Going from a kid who watched the American icon on our black-and-white TV in the '50s to having complete access to him and his office—well, let's just say it was one of the proudest times of my professional life.

Bringing in fresh ideas for television productions, with the emphasis on stories that came from news pieces and other concepts was my job. It was fun to watch the process play out. I, or one of the other half-dozen associates, would walk through the doors of his office and pitch an idea. If Dick was interested, he would get up and walk across the street to NBC studios to pitch it to the network executives. Being Dick Clark gave him direct access to the ones who could give the OK to new projects for the network. If he didn't like the idea, he stayed put behind his desk. In 1991, I had the good fortune to have an idea that sent Dick jogging to NBC.

Word got around after I had a few successes under my belt in developing news pieces into media pieces. Everyone approached me with ideas for stories; it got to be a little annoying at times. If I was picking up my kids from school or having dinner at a restaurant, I would often hear, "Hey, Larry, I have this great idea for a show!" I even remember sitting in a doctor's office waiting for one of my kids when the receptionist approached me and pitched her idea for a show based on her experiences when pesticides contaminated her house. In the early 1990s, a couple of seedy-looking characters walked into my office saying that they had videos of people breaking the law. At first, I thought maybe I should call security; but when they explained that they were bounty hunters and had videos of their experiences catching bail jumpers, I started to listen. They wanted us to produce a show about them; but after thinking about it for a while, I thought about having a show that would not be limited to people who were running from the law. Why not produce a show that would catch people from all walks of life in a variety of situations instead?

This was before George Holliday videotaped the beating of Glen King by several LAPD cops. The first reports got King's first name wrong. Instead of using his real name, Glen, the news reports mistakenly reported that his name was Rodney. The name stuck, and he has been known as Rodney King ever since.

Before the Rodney King tape made its mark in history, news programs aired real-time videos of crimes occasionally, and it was uncharted territory as far as television entertainment programming was concerned. But, I thought scenes of people committing crimes could capture the attention of the viewing public and be a force in the ratings war.

Research found that many police departments and the FBI could make videotapes of criminals committing a variety of crimes available to us. Sometimes it was simple shoplifting captured on in-store surveillance cameras, or a video sting set up by law enforcement to have proof to convict a drug dealer in court. With some of the videos in hand and a brief outline of how the show could be formatted, I approached Dick Clark. It usually took a couple of hours to go over the complete plan and outline of a show; this one took less than thirty minutes. At the end of our short meeting, Dick was rushing out the door on his way to NBC, and I was heading back to my office to think of someone who could write the script and a star to MC the show.

At the Dick Clark Film Group, we had an environment that was easy to work and create in. Dick had an open-door policy, so whenever we had ideas or needed his input on a project, it was just a matter of walking into his office and pitching. Even the physical surroundings were conducive to creating television and film entertainment. He had purchased a building that was an old motel, then converted it to Hollywood-style bungalows. So if the producers were burning the midnight oil and needed privacy, they had the old Hollywood setting. We also had the power of being able to contact just about anyone we wanted. It didn't matter if the target was a bus driver or the White House chief of staff; with a little creativity and a lot of persistence, I could get to anyone. It was a great lesson for me in both the news and entertainment sides of the business.

After accumulating a variety of different videotapes of people breaking a wide range of laws, the screenwriter started on the script. We were fortunate Dean Stockwell, who was well-known for his role on the television series *Quantum Leap*, agreed to be our host. We watched hundreds of tapes; some boring, some outlandish. One that sticks out in my mind was a tape that caught an average-looking housewife parking her station wagon full of kids outside a seedy apartment. She left her young children in the car while she went inside to perform sexual acts in exchange for drugs. Dick and I knew it was a little much, so we did not use it in the pilot. We had plenty to choose from anyway and ended up with a variety of shoplifters, people committing insurance fraud, casino cheats, and dirty politicians getting payola. We had the most fun with segments called Stupid Criminals. I knew the public would be enticed by reality.

As the impending airdate neared, we decided to call the show *Caught in the Act*. It seemed appropriate, and the description was as straightforward as it gets. We also put a tiny camera on Dean's lapel without the audience knowing to show how easy it really was to be caught on video.

With the risk of sounding cliché, Dick and I were sitting in director chairs with Executive Producer and our names silkscreened on the back watching the process come together. Dean delivered the script without a hitch, and the bit with the hidden camera in his lapel was intriguing. The opening dialog said it all, "The stories you are about to see are real. Only the scenes outside of the surveillance footage were created for dramatic effect." Dean went on to say how crime was on an uncontrollable rise, from the guy who fakes a fall for insurance money to bank robbers. About halfway through the taping of the pilot, Dick leaned over and whispered in my ear, "You know Larry, I thought about this years ago." I was on a roll and had my boss sitting next to me telling me that he had the same idea years before but never acted on it, and here we were filming a pilot from my creative efforts. I couldn't help myself. I leaned over and whispered back into his ear, "So why didn't you do it?" We smiled. We both really felt that we were on to an idea that could capture the ratings. Little did we know it would be the wave of the future for reality television and network news.

Surfing the channels now during prime time, it is impossible not to see a reality show. The American public has demonstrated its voyeuristic desire to look over peoples' shoulders while they live their lives. I am not going to go so far as to claim that I had anything to do with the popularity of reality TV, but I do see how the shows that capture your attention today can be traced to an idea that Dick Clark and I developed years ago. In the old days, network programmers looked at news and entertainment as two separate entities. The shows that currently capture the numbers in the ratings war illustrate the blur between the two.

Viewing audiences are no longer satisfied with watching actors practice their craft from rehearsed scripts. Even though the "reality" in reality TV is questionable, the public has shown that it is interested in watching situations unfold that are not the product of a screenwriter's imagination. The ratings convey that people are more interested in watching what it is like to be stranded on a deserted island where your fellow castaways can vote you off to win a million bucks, or have a multimillionaire tell some poor crony that he is fired, or see how a real-life girl is going to choose a man from a pool of several dozen than in seeing anything imagined. If I had presented ideas like this before to the networks, I would have been laughed off the studio property. Now the major networks realize that the public is hungry for the real stuff. Ironically, I did pitch the bounty hunter concept to the networks after doing *Caught in the Act,* but no one was interested then.

The news media has also taken notice of the public's craving to watch the real-life videos of people doing things out of the ordinary. ABC and all the other news agencies have little trailers run at the end of their respective newscast that tell viewers how to contact them if they have any home video that is newsworthy. Even former vice president, and almost president, Al Gore, has taken advantage of the video boom. His cable network, Current, relies heavily on amateur-produced videos, not unlike what we used for *Caught in the Act.* When we filmed our show, the videos came mostly from sting operations and the like. Now, technology makes it possible for the average person to own a video camera portable enough to be handy and still have the quality needed to be broadcast worthy. A few hundred bucks gives anyone the ability to go out and hunt down news stories and capture the images on video, and everyone in the industry knows it. I never leave home without my news media pass, a camera, a video camera, and my cell phone that has both. Being in the news business, I know that a newsworthy story can happen anytime, anywhere, and I want to be prepared.

The argument for the right to privacy has been almost muffled by the argument that we need to be more secure. By keeping us all under surveillance we can accomplish just that, say the people who want to keep the cameras trained on us wherever we may be. I can't help but to wonder if the ideas we brought to the networks may have accelerated the process. The only real difference between what we did and what the networks are doing with reality shows is that the people being filmed know the cameras are there.

Caught in the Act opened my eyes to the potential of having news stories big and small receive the attention of producers from the entertainment side of the networks. While producing the pilot, it was impossible to ignore how fascinated the entertainment guys at the network were. They were eating up the concept.

From experience gained in developing the pilot and placing news stories, I learned that I could make a profit from stories I had acquired that I had given away before for free to the news shows. News agencies were willing to pay—in a creative way, of course—to be able to put some of the news pieces on the air. So, instead of reporting my stories with a spin, I produced them and told the truth for my clients, which gave me the ability to reinvest any producing fees made into the development of a film and/or a book. I was not working one against the other, I was working both together; and it has been very successful for my clients and me.

Being involved with *Caught in the Act* was a turning point in my career, but I still feel like we may have created a monster. Rodney King's video shows just how destructive it can be. If there had not been a video of his beating, it probably would have been a quick mention on the local evening news. But because home video captured images that the networks could play over and over, it made the story bigger. People did not have to rely only on what the reports said. They could watch the events unfold and make judgments for themselves. A couple of years after the video was first aired, the officers involved went on trial and were found not guilty of any wrongdoing in the beating of Rodney King. A jury of twelve had the added benefit of hearing and seeing evidence besides the video to form the opinion that the officers involved acted correctly. The problem is that the public did not have all the information, only the video, which enraged the black community to the point that riots sparked in Los Angeles shortly after the verdict was announced, killing dozens of people. The video also drew in the entire nation, and smaller riots occurred in other large cities across the U.S.

On the surface, it would appear that video of an act would provide irrefutable evidence, so that a conclusion of what happened could be formed with no room for speculation. Even video of things that happen after the fact helps us form opinions relating to certain cases. If O. J. Simpson had not been filmed being chased by the LAPD in the white Bronco before his arrest for the murders of his wife and her friend, maybe the public wouldn't have been so quick to assume his guilt—even after his not-guilty verdict was read before a huge nationwide audience. Like I always say, there are two or more sides to every story, and then there is the truth, even if part of the story has been caught on videotape for us all to watch. My fear is that if the public's desire to have instant answers is satiated by surveillance tapes, it might relieve the news agencies of their responsibility to verify every part of a story. No matter how you look at it, with the influx of new videos that capture the news and the public's tendency to watch things that they consider to be "reality," whether on the news or for their entertainment, the face of television has changed. We have become a nation that loves to look into the lives of others and are quick to pass judgment on them from the glimpse of their lives that we see. But we still resent having our own privacy violated.

It has made TV news and entertainment more complicated, but Big Brother is here to stay. The privacy we used to take for granted is gone, and the ratings wars will help to ensure that the tools of surveillance, like undercover cameras and recorders, will always be operating silently in our day-to-day lives.

It has also added to job security for others and me who procure an individual's news rights. The news rights we obtain often include videos as part of the story, and I can only hope that the news media will be responsible enough to 1) never stop seeking to tell the entire story and allow undercover videos to do their job for them; and 2) never allow this new tool of spin to control the media monster.

9

Making a Difference

When I first meet people who know about my involvement in the news process, I can always predict the questions that will inevitably be asked. "What was Larry King really like?" "Do you think Blake murdered his wife?" "Is Michael Jackson a pedophile?" "Tell me, do you think Natalee Holloway is a sex slave, or what really happened?" "What was Andrew Cunanan's family like, and did they hate Gianni Versace?" "What was it like at Heaven's Gate?" "Tell me more about Columbine." And the rambling has just begun. "Why did you expose Jerry Springer on *20/20* for staging his shows? Was it because you caught your kids watching him?" "Does General Thomas P. Stafford [commander of the moon landing, who also helped to bring down the Iron Curtain] believe in UFOs?" "Give me the scoop on the White House." Etc., etc., etc.

I understand the curiosity and the desire to get the inside scoop on some of the bigger stories; it is what drives the news business. I always field the questions I am comfortable answering and politely dodge the others for one reason or another. All the while, though, I wish they would want to hear about the stories I am most proud of—the ones that have made a difference.

In my work, I have seen and spoken to a wide range of personalities. For the most part, the people with the stories try to find an angle that might allow them to put a few bucks into their pocketbooks. But then there are the exceptions. People who have a high level of integrity and are willing to put themselves though a great deal of risk and personal pain to reveal a truth. Their motivation oftentimes is only to make sure the truth is told. Sometimes it is a situation that has the potential to harm many, and other times it is someone who has been in a position of public trust and has misused that trust. Once in a great while, I happen across someone who is willing to endure public scrutiny and relive the pain and humiliation of a horrible event. These are the people who truly make a difference, and I am proud to be associated with them. Courage, compassion, strength—these are just a few of the words that can be used to describe these individuals who ignore their fear of the media monster and forge ahead to make something good come out of a horrible situation. The first person that comes to mind when I think of this type of hero, who fits into this rare category, is Lisa Celestin. Her unselfish actions made the world a safer place for our daughters by bringing to the public and to Congress the reality of a most damaging crime.

Most crimes have lines that we don't cross. There is little or no room for speculation. When someone steals something from someone else, there is no question that the act was wrong and should be punished. The public will not accept any justification for acts like murder, or when individuals claim to represent the victims of Hurricane Katrina but pocket the donations for themselves. The line that dictates what is right and what is wrong is clear. Sadly, one of the most brutal and devastating crimes has lines that are often blurred, even though it is one of the most damaging intrusions into an individual's life.

Rape is a crime that leaves the victim scarred deeper than

any knife can penetrate and in a position where they are on the defensive. Feelings of embarrassment and humiliation invade the victims because of the process they have to endure to report the crime. Having to explain to a complete stranger what happened to her, subject herself to a battery of intrusive physical tests, and field questions that make her relive an event she would rather put behind her is a very difficult situation. Over the years, the public's tolerance of violence toward women has, thankfully, diminished. But it does not make it any easier for a victim to seek justice.

Studies of the statistics vary, but most agree that approximately 40 percent of sexual assaults on women, including rape and attempted rape, are never reported. The stats show that it is difficult for women to even go through the process, and that emotional scars prevent almost half of the victims from seeking justice.

Currently the number of rapes committed is on the decrease. Experts suggest several reasons for the decrease, but I know that a big part of it is because of the courage that Lisa Celestin and women like her displayed as they stood up for their rights as human beings and said enough is enough. In Lisa's case, she not only exposed the crime of rape and its true nature, she also ensured that the use of a particularly sadistic tool in the commission of that crime would carry a much stiffer penalty.

Until the 1980s, alcohol was the weapon of choice for men who wanted to violate women. But a new drug was developed for a much different purpose that armed potential rapists with a tool that rendered women unable to defend themselves, and many times robbed them of the memory that they had been violated in the first place.

Swiss pharmaceutical giant Hoffman-LaRoche developed Flunitrazepam as a sedative in the early 1970s. Valium (also developed by Hoffman-LaRoche) made its mark in the 1960s as the often prescribed "feel-good drug" that reduced anxiety. One of the nicknames for the pill was "Mother's Little Helper," as it was often prescribed to baby-boom-era women who needed that little bit of relief from the challenges of motherhood. It was also the inspiration for the Rolling Stones hit song of the same name in 1966.

By the 1970s, Valium was prescribed as a lifestyle drug, and the line between therapeutic and recreational use became blurred. Although it was legally prescribed by doctors of the day, it became apparent that the drug was being misused and was highly addictive. Like Valium, Flunitrazepam was effective in treating muscle spasms and seizures, but it was much more powerful. It took about an hour for the effects of Valium to take hold; with Flunitrazepam, effects were felt in fifteen to twenty minutes and were as much as ten times greater. The effects of the new drug also last four to six hours, with residual effects lasting up to twelve.

Flunitrazepam's trade name is Rohypnol. It has never been approved for use in the United States but is, nonetheless, readily available and used recreationally in the country. At high school and college parties, the drug is used to produce profound intoxication. Cocaine and heroin users call a dose of Rohypnol a "roofie" and dangerously mix the two to achieve a better high. As the popularity of the drug increased in the United States, so did the amount of reported rapes, and a new term became part of our vocabulary: date rape.

Women reported that they suddenly and inexplicably

became incapacitated to the degree that they had little or no memory of what had happened to them for hours. Somewhere between life and death, there is a state where a person is not really awake but not really asleep. It is an altered state where a woman becomes vulnerable and finds herself at the mercy of whoever's company she is keeping.

Unfortunately, many men—I should say *creeps*—purposely administer the drug by slipping it into women's drinks without them knowing, then wait for the effects to take hold. After fifteen to twenty minutes, the woman becomes completely helpless, in a state of physical paralysis, and he commits the hideous crime. What makes things even worse is that in many cases Rohypnol robs the woman's memory. There may be a fuzzy recollection of what happened, but not enough to file criminal charges against the rapist who violated her, let alone have a case that could stand up in a court of law. In short, when a man slips a dose of Rohypnol into a woman's drink, he takes away her ability to say no or to defend herself, and leaves her with only a foggy recollection of what happened. It is premeditated. He has a plan in mind.

What is equally concerning is that it is an easy crime to commit. The drug comes in tablet form and can be crushed so it quickly dissolves in a drink, and it is also available in liquid form. It has no odor or taste, and only when the effects of the drug start to take hold does a woman know she might be in trouble. By then it is too late.

If she remembers at all, it is usually the day after the rape that a woman realizes she has been violated; but memories of the act itself, as well as the events that led up to it, may only be a blur. A kind of cruel amnesia takes hold. The confusion and uncertainty can cause a woman to delay reporting the crime, or sadly, not report it at all. For those who do report the rape, it is often too late to gather physical evidence that she was drugged in the first place. Rohypnol can be traced in a urine sample for only seventy-two hours. With each passing hour after the drug is ingested it becomes harder and harder to detect, so even a slight delay in testing for the drug can make it impossible to know for certain if the drug was used or not.

Innocent until proven guilty is the standard of measure. The only way to prove guilt in a rape case is either physical evidence or reliable witness testimony. With Rohypnol there is neither. The residue from the drug passes through the body quickly and is often not detected, and the memory loss it inflicts on the victims can cause their testimony to be viewed as unreliable. There may be evidence that a sexual act occurred, but when the victim cannot remember what happened, it is easy for the defense to argue that it was consensual sex and not rape.

As a father of two daughters, I want to do what I can to help my girls be safe. In 1995, I was scanning the wires when Lisa's story caught my eye. It wasn't a big story, just the mention of a woman who had accused a man of raping her and using Rohypnol to commit the crime. I contacted her attorney to see if she would be interested in getting her story more attention. In the days that followed, I spoke with Lisa and got a much clearer picture of how devastating the crime was. I was outraged and saw the importance of helping her get her story out.

Lisa was out with some friends socializing at a local nightclub. Everything seemed normal, and she was just having fun when she began to feel a little confused. It took only moments for the Rohypnol to take its full effect on her, and within minutes she completely passed out. It wasn't until the next day that she realized she had been raped. The rest of the night after the Rohypnol was slipped into her drink was just a void. She had scattered memories, including a moment when she woke up while the man was raping her. Perhaps her biggest frustration was that she could not remember enough about the night of the crime to press criminal charges against her attacker, even though he laughed in her face a few days later as he bragged about how he committed the crime. He drugged Lisa to the point where she was incoherent, then drove her to her house and raped her in her own bed. There was absolutely nothing she could do about it while it was happening or afterward. The word *violation* comes to mind when trying to describe the pain she was going through, but it does not come close to describing the emotional scars.

Instead of hiding from what happened, Lisa decided to speak out. She wanted to use what had happened to her to raise public awareness of the risk to women because of the date-rape drug. Lisa was a thirty-five-year-old mother of three, living a fairly normal life in Florida until the night of the rape. Then her life changed at the hands of a stranger. She was in fear of AIDS and other sexually transmitted diseases, and she was let down by a system that exposes women to the threat of being raped in this manner. When she learned that the law did not impose severe penalties for men who used Rohypnol as a weapon in rape, she made it her mission to see that things changed.

Raising public awareness was our intent when I agreed to help Lisa, but it was not as easy as I had anticipated. The news agencies were not eager to cover the story. If I had come to them with celebrity gossip or behind the scenes information about the O. J. Simpson case, I would have been welcomed with open arms; but when it came to the subject of rape, the doors were closed. It seemed to be one of those subjects that everyone in the industry knew about but didn't want to devote airtime to. After a few persistent weeks, I finally got a break and got Lisa on the news. I parlayed this success into many other interviews on the major networks, and people started to notice. Because of the courage that Lisa displayed during the interviews, people started to realize that the problem should not be swept under the rug but should be talked about openly, and action should be taken to discourage it from happening again. I noticed that other women were also starting to speak out about their experiences, and public awareness of the use of Rohypnol in the commission of date rape was on the rise.

Programmers let me know right away that there was not a chance in hell of anyone watching a movie about date rape. In the early 1990s, network entertainment directors were not as liberal as they are now, and any proposal that even touched on a sensitive social issue like rape was flatly rejected. I had hoped a movie could be done to increase awareness, and I felt Lisa deserved to be compensated for what she was doing—but it wasn't going to happen. That did not deter us from pursuing the news outlets to cover the subject, and the government noticed.

About a decade before Lisa was raped, Rohypnol was placed on the controlled-substance list as a Schedule IV drug. The official definition published in the Controlled Substances Act of a Schedule IV drug is as follows:

> The drug has a low potential for abuse relative to other drugs, is currently accepted for medical use in treatment in the United States, and may lead to limited physical dependence or psychological dependence relative to drugs in Schedule III.

Rohypnol was not legal in the U.S. but was acceptable for medical use according to the Controlled Substance Act. The higher the number on the Schedule, the less risk it has in the eyes of the law. There really were not any consequences for people who brought it into the country, sold it, or used the drug for themselves or in the commission of a crime. Rohypnol was grouped with drugs like Xanax, a widely prescribed mild anti-anxiety medicine, and Valium. Law enforcement generally focuses its attentions on drugs that fall in Schedule I and II, drugs like cocaine and heroin. If someone was caught with Rohypnol, it was kind of like breaking the speed limit. You were in trouble, just not very much.

Lisa Celestin, along with Joy Diliello—also a victim of a date rape involving the use of Rohypnol, who became pregnant because of it—testified before Congress on July 17, 1996. Lisa and Joy were not attorneys or politicians, they were everyday average women who had suffered a great deal at the hands of men who used a drug to rape them. Their testimonies were heart-wrenching. Lisa kept her cool and was very deliberate. She described that she couldn't remember any of the events of that night. "This guy could have sawed me in half and I wouldn't have known the difference," she said.

The fact that Congress was considering reclassifying Rohypnol was a major victory. News coverage increased and public opinion of date rape went from seeing it as a high school prank to viewing it as a horrible crime that deserves to be punished severely, and that steps were needed to prevent it. The attention the hearing brought also motivated the drug's manufacturer to reformulate it so that a drink would turn bright blue if spiked with Rohypnol. Although there is still a supply of the older pills that stay clear when dissolved, and creeps who will serve blue drinks to mask the new pills, it is a step in the right direction.

On the federal level, Rohypnol is still a Schedule IV drug, but with the publicity and actions that Lisa and others like her took, many states have reclassified the drug to Schedule I. In those states, the possession, use, and distribution of Rohypnol carry the same penalties as do heroin, LSD, and methamphetamines. In a majority of the states that have not reclassified the drug, there are bills in the works. So, we didn't win the war on the federal level, but many battles are being won on the state level to make the chances of a woman being violated with the use of Rohypnol much smaller.

Looking back at what Lisa did and the effects her actions had, I am most proud of the fact that she helped raise public awareness of a horrible situation regarding the attitude toward the crime of rape in general. Reported cases of rape are down, and I believe it is partly because of the courageous actions Lisa took. The doors were opened so public discussion could begin on a subject that was considered taboo by the news agencies and entertainment industry. The understanding that society as a whole will not tolerate crimes of this nature, and that many states will lock a man up for twenty-plus years has had a positive effect on the statistics on all types of sexual assault.

The date-rape drug would come back to haunt me. After

working with Dave Holloway on the infamous disappearance of his daughter, Natalee, I realize that drugs like these are not adequately controlled on an international level. As we detail in our book *Aruba*, there were reports and rumors that Natalee might have been drugged the night of her disappearance. Until we know Natalee's fate, we may never know the truth of that grim night, but I do know that tourists and travelers must be very vigilant.

In a day when TV news devotes much of its attention to sensationalized stories, it is good to know that a small voice can make itself heard and have a positive impact on our world. Lisa was not afraid to speak out and be persistent; to make sure her voice was heard and in the process make society a little safer for our sisters, wives, and daughters. She truly made the difference.

People do not have to be victims of crime or be willing to speak out about injustice to be in a position to make a positive change in society. Changes can be subtle and still be just as important. Sometimes, someone comes along who is special and the love of life and willingness to live it to the fullest, no matter what the challenges, are impossible to ignore. In fact, their actions prove to be inspirational and can change the way we look at things. In the early 1980s, I had the privilege of being involved in a story about a little girl whose infectious smile and beautiful attitude about life made a remarkable difference for me and many more.

Tracy Taylor was the poster girl for the March of Dimes, the charity founded in 1938 by President Franklin D. Roosevelt to fight polio. Initially, the charity was called the National Foundation for Infantile Paralysis. Roosevelt wanted to make sure that the charity was nonpartisan and public in nature, something to appeal to the masses and not just the rich. It began with a radio broadcast, urging everyone in the country to contribute a dime for research to find a cure for the disease that affected thousands of Americans in the first half of the twentieth century. In 1938, the public received much of its news via the newsreels that played at local movie houses. One of the more popular newsreels was *The March of Time*. Entertainer Eddie Cantor used a play on words when he contacted one of the fundraisers, and soon the charity became known as the March of Dimes. In 1979, it became the official name of the organization. The March of Dimes has become a major force in raising funds to fight a multitude of childhood diseases. Roosevelt's close affiliation with the charity is the reason he is portrayed on the U.S. dime today.

Originally, my involvement with Tracy was not only to develop a TV movie about her life. I knew Tracy's story was important for providing hope for kids with disabilities. I knew firsthand of the challenges she faced. Many of my friends had battled the disease. In 1954, I was part of the experiments with Dr. Jonas Salk that later proved to be the end of polio. I still pull out the little button they gave me years ago to remind me of that challenging time. I would have worked to get her story out for nothing in return.

I had partnered with Dick Clark and we wanted to tell her story to show the audience that people with disabilities were still people. She was an incredible little girl. Although she needed crutches to walk, Tracy's attitude about life was that it was too precious to waste. She was fun-loving and wouldn't let her disability stop her from doing things that many in her condition felt were impossible. She had learned how to roller-skate and even did gymnastics on horseback. In fact, Warren Miller, the well-known guru of ski movies, asked me to let him use her in one of his movies and taught Tracy how to snow ski in a week's time.

Changing the public's view on the disabled was our goal, and Tracy's story was a perfect illustration of why this was important. In the '70s and '80s, things were different for people with handicaps than they are today. Before the public was made aware of the challenges faced by those with disabilities, many public buildings made it all but impossible for someone who could not walk to gain access. Restrooms were impossible for many to use. The overall attitude about someone with a disability being able to live a full and complete life was that it didn't quite seem possible.

When we presented the idea to the major networks, we were turned away. It seemed that the entertainment directors shared the lack of understanding that the public had about the disabled. Dick and I heard things like, "We aren't doing anymore disease of the week stories this year," or "You need to focus on more timely issues." That was the first time I had heard "the disease of the week," crack. It was disappointing and sickening to me.

Tracy's spirit was infectious and kept me focused. I had to remind myself that this young girl had lost her mother and had been inflicted with a disease that left her disabled. I knew that her story could be an inspiration to many—not only for those who were handicapped, but also to those who faced challenges as the result of setbacks they had suffered. Even though we could not put together the TV movie deal, I decided that her story was too important not to get at least some of the attention it deserved. So, I went to the news agencies.

At first, I met the same resistance I had encountered with the entertainment directors. The news programmers didn't see how her story was important enough to relinquish any of their precious airtime. At that time, they felt that the public did not want to hear about how a handicapped little girl was living an incredible life despite her disability. There were a couple of times I was almost ready to give up, but I thought better of it. The story itself wouldn't let me. How could I not try to overcome the obstacles I faced when Tracy refused to stop trying? This little girl really proved to be an inspiration to me.

People magazine was convinced to run a story on Tracy called "A Child of Joy," which introduced the general public to Tracy, then we went to the news agencies to run pieces on her. It took awhile, but we started to get her story told on some of the news shows. People got to see that the excuses in their lives were trivial when they saw what a ten-year-old girl was overcoming.

It is disappointing that a movie deal did not come together for her. But, I can see the impact her story did have. When I go to the movies and see areas reserved for people in wheelchairs or to restaurants and walk up ramps instead of stairs, I realize that Tracy and people like her did make a difference. Theirs are the stories that raised public awareness and motivated the changes necessary to give people with disabilities the same access to public events and places as the rest of us. Even the overall attitudes about the disabled have changed. The handicapped are not treated like second-class citizens anymore. They receive the same respect and rights as those who do not have handicaps. This is one of the times I am proud to be associated with the news industry. Instead of focusing on the negative and airing only what the industry thinks you want to see, a few of the outlets were willing to tell Tracy's story and others like it. It may not have been the best move in terms of the ratings wars, but the media monster did bring to the public's attention stories that led to good things and improved lives.

It is a privilege to be part of stories that have a positive impact. I fully understand the business and how it operates; business is business. But, long after I am gone and the sensational headlines are a distant memory, these stories will continue to make a difference in our society. This makes my work rewarding.

10

Should Have Been

Being involved with some of the biggest stories to capture the headlines over the years has been thrilling. Switching on the TV set and watching the news pieces I helped bring to network news gives me a great sense of accomplishment. It is a win, win, win situation. The networks have the ammunition they need to stay competitive in the ratings wars, my clients often get their fifteen minutes of fame that can lead to other successes in their lives, and I enjoy a professional satisfaction that is often reward in itself.

But then there are the stories that I truly believe should have captured the headlines and been given the potential to have a positive impact on the public, but for one reason or another did not get the news coverage they should have. There are many. The first one that comes to mind is a group of doctors in Chicago that I became involved with that believed they had found the cure to the plague of the twentieth and twenty-first centuries. They were so convinced of the worth of their discovery that they were willing to inject the plague into their own bodies to prove their point.

In 1981, there was no shortage of big news stories to capture your attention. Ronald Reagan was sworn in as America's 40th president, and Iran decided it was better to release the fifty-two hostages they had been holding captive for 444 days than to risk the Gipper's fury. Across the Atlantic, English royalty captivated viewers as Prince Charles and Lady Diana Spencer wed in the wee hours of the morning, U.S. time. Air-traffic controllers had a plan to increase their salaries by trying to freeze air travel in a nationwide strike. Reagan altered their plans by firing them. One hundred and eleven people died tragically at the Kansas City Hyatt Regency when an overhead walkway suddenly collapsed. The players in the National Baseball League went on strike for seven weeks. America returned to space when the space shuttle *Columbia* completed its first successful mission. The country's first test-tube baby was born three years after the world's first in Britain, and the pope and the American president survived being shot in assassination attempts. It was a fairly eventful news year, even by network standards.

Surfacing, in 1981, was also a new sickness, although the public in general would not hear about it for a few more years. It induced in its victims a very rare form of skin cancer and attacked their lungs with an even rarer type of pneumonia. Doctors were puzzled, as patients who exhibited the symptoms did not respond to the treatments that had worked before. And they died. As medical professionals looked at the patients closer, they found that their immune systems had been destroyed by the disease, and started to realize how cruel and deadly it really was. Not only did it make people sick, it also robbed the body of the ability to protect itself and get better.

Centers for Disease Control started to look into the new disease that proved to be so deadly. In the U.S., many of the victims could trace the origin of their disease directly or indirectly to a Canadian flight attendant named Gaetan Dugas. As with the vast majority of the people who were infected, Dugas was gay, and the sickness was transmitted sexually. Because of the high concentration of gays in the population who had contracted the disease, it was named GRID, short for Gay Related Immune Disorder. The new sickness got a few mentions on the news, but for the most part the American public was completely unaware of it. As time passed, other parts of the population started to exhibit the same symptoms, and the disease began to spread in the general population. Drug addicts that used intravenous needles and people who had blood transfusions were also succumbing to the new disease, and many of the victims swore that they were completely heterosexual. Then children became infected and it was clear that the new disease was not confined to only the homosexual population, so it was renamed AIDS, Acquired Immune Deficiency Syndrome.

Scientists on both sides of the Atlantic were studying the new public health threat, and in 1983 scientists in Paris at the Pasteur Institute isolated a virus that was linked to AIDS. In 1984, American scientists also isolated a virus that they believed was the cause of AIDS. Two years later, the new virus had a name, HIV, or Human Immunodeficiency Virus, when it was discovered that the virus could be transmitted person-to-person by contact with HIV-infected blood.

AIDS was gaining recognition as news agencies started to cover the disease and as celebrities became victims. Rock Hudson was one of the first major celebrities to die from AIDS, and in 1987 Liberace was also a victim. Other well-knowns to die from AIDS included tennis great Arthur Ashe, world-renowned Russian ballet star Rudolf Nureyev, *Brady Bunch* dad Robert Reed, and the lead singer for the rock group Queen,

Freddie Mercury. Basketball star Magic Johnson announced that he was HIV positive in 1991. With the death and infection of the famous and the far-reaching implications for the world population, AIDS became one of the top news stories in the late 1980s and early 1990s. In 1987, thirteen-year-old Ryan White made headlines when he was not allowed to go to school because he had AIDS. The fatal disease rapidly escalated from 152 reported cases in the U.S. in 1981 to an estimated 2.8 million worldwide in 1999.

In 1984, U.S. Health and Human Services Secretary Margaret Heckler announced that a blood test would be available to screen those infected with HIV and that she was confident a cure would be developed within two years. The worldwide medical community frantically tried to find a cure, and in 1987, a new drug was developed that showed great promise, at least at first. AZT (Azidothymidine) seemed to perform miracles in that it all but stopped the progress of the disease, but after a year, it proved to be only a temporary miracle. Patients started to develop a resistance to the new medicine, allowing the AIDS sickness to continue its destructive course in their bodies. But the major drug companies still had a cash cow in AZT, as it was the only drug that provided relief—if only temporarily.

In January of 1998, I happened across an Associated Press article that covered a group of doctors that believed they had made great progress in finding a cure for AIDS. When I read that they were willing to inject HIV into their own bodies to prove their point, I immediately contacted the group.

José M. Zuniga was the deputy director of the International Association of Physicians in AIDS Care. The group, based in

Chicago, represented more than 5,500 physicians and health-care professionals who were dedicated to finding a cure, not just a treatment, for AIDS. After our first conversation, I quickly understood their challenges.

Rumors pervade about companies that control discoveries or technologies to protect their financial interests. One of the popular ones is that the oil companies buy up inventions that would greatly increase the gas mileage of cars or provide alternative sources of energy. These rumors have been around for years and may or may not be true. In the medical industry, there really are some major hurdles. What José explained to me was alarming. His group of doctors had been working tirelessly on a cure for AIDS. What they found to be frustrating was the politics involved in getting their information to the public, even though they had made a great deal of progress in their research.

Battling bureaucratic red tape and lobbyists from the major drug companies hindered them from having their discoveries published. The drug companies in this country are big, big business, and literally billions of dollars are involved in the industry. With this much financial clout comes powerful political players that can stand in the way of cures being made available to the American and worldwide public. Obviously, the company that is first to develop a cure for AIDS stands to make an enormous amount of money. With the lure of those potential profits, the big boys would like to muscle out any competition from a smaller group that might develop a cure that would affect millions.

The hurdle the International Association of Physicians in AIDS Care was confronted with was that the government and individuals in the medical field did not want to take them seriously when they claimed they had made substantial advances in finding a cure. Their motivation was the fact that in every day that passes without a cure, sixteen thousand more people become infected, including two thousand children. His group was, and is, dedicated solely to the cause on a human level, not a financial one.

José also explained that the group's efforts to get the word out about their discoveries and attain government support to aid in the promising research was getting nowhere. They were met with obstacles of the medical community wanting to focus only on "pure science," instead of bringing a cure to the masses quickly. His group consisted of people respected in the medical community, not a bunch of publicity seekers. They truly felt that the only chance to get the attention and funding they needed to pursue their research was to make public their intention to inject a weakened strain of the HIV into their own bodies if they had to, to expedite the discovery of a cure.

It may have sounded like a publicity stunt, but these people were serious. They believed so much in their research that many of them were willing to risk their lives to get their point across. The thought of doctors who were family people, being willing to put at risk their reputations, health insurance, and lives by injecting a strain of HIV into their bodies was overwhelming to me.

Interacting with José and his associates made me realize that if the general public knew of the group's dedication and advancements in the research for a cure, the chances for their work to be taken seriously by the medical community and government would increase. Maybe it would bring a cure faster and save lives.

Money is the issue, like anything else of this magnitude. I agreed to help the group get more attention in the media and to use my resources in the entertainment and music industry to raise funds to aid in their research, and help them make the public aware of their intentions. Hopefully, this would put pressure on the government to take them more seriously. We had some success, but it was not easy.

The AIDS epidemic was like the date-rape issue in the early 1990s as far as getting the media to cover it. The news agencies were hesitant about covering groups or individuals outside the mainstream medical and drug companies, even though the news directors knew that the research José's group was doing was credible. Unless I could wave documentation that a cure had been discovered and verified by the American Medical Association and the FDA, the major networks shied away from dedicating any airtime. Although I wish I could have drummed up more interest in the news, I did manage to get the group on a few shows and gave them a chance to get the word out about their research. It was only a mild success when looking at the overall picture. The group did get a little publicity and received more funding from the private sector, and people in Washington did take them a little more seriously; but overall I was disappointed. With thousands of people dying every day from AIDS, and mainstream companies failing to develop a cure, I felt it was important that other avenues be explored. A mother whose son may be cured of AIDS doesn't care if a multibillion-dollar drug company developed the cure or if it was discovered in the kitchen of a dedicated health professional. It's not important who finds the cure, it's that a cure be found quickly.

The International Association of Physicians in AIDS Care is

still relentlessly researching the cure and even with their limited resources, is making progress. At this time, no one in the group has injected HIV into his or her own body, but many in the group still entertain the thought, if it would make finding a cure faster. I wouldn't be at all surprised if one day soon I will be helping the group put together a press release stating that several of their members did inject themselves with HIV in hopes of finding a cure. Maybe then they will get the attention and funding they need to help get the job done. Although I am not ready to admit defeat as far as getting the network news agencies to take notice of this group, and others like them, I am frustrated that people may be dying needlessly because of a system that prevents people who think outside of the box from being heard.

Fantastic stories that could have impacted a large segment of the population, like José's group and their quest to find a cure for AIDS, should have received more attention from the media, but sometimes something will slip through the cracks that affects only one person. Although the impact is not the same as the news pieces that affect the masses, many times the story is important and should be told.

In 1997, I procured the news rights to a man named John Futrell. Although you may have never heard his name before, you probably have seen his picture; it was printed in every major newspaper and aired on every news show I can think of. The image of a police officer kneeling over a handcuffed man on the ground told only part of a disturbing story.

Visitors to the area of North Hollywood would see an area densely populated with rows of one- and two-story commercial buildings that crowd the grids of major streets. Between the big grids are the smaller streets that hold houses and apartments. On the commercial streets is the normal array of small businesses—liquor stores, ma-and-pa businesses, gas stations, doctor and dentist office complexes—mixed with some of the larger chain stores and banks. On February 28, 1997, two masked, heavily armed men were robbing the Bank of America at 6600 Laurel Canyon Boulevard. Someone inside was able to send out the alarm just after 9:00 a.m. and alerted LAPD to the 211 in progress. Squad cars were immediately en route to catch the bank robbers.

Initially, it was reported that up to six men were involved in robbing the Bank of America, and with what was about to happen, it might have been less embarrassing for the LAPD if there had been six; but it was only two. These two men, Larry Phillips and Emil Matasareanu, were armed with a trunk load of automatic assault weapons, and they wore body armor. This situation would have played out like most other robbery attempts that fail—the suspects either surrender and are arrested when they see that they are surrounded by police, or they are shot by a police bullet and sent to the morgue—except that within moments of these two seeing the first police car, they began firing, shattering the peaceful quiet of the morning. Police radios were filled with screams from officers describing the two men randomly firing into the area and how they were helpless to stop them. I know firsthand what it was like to be within earshot of the start of the shootout, as I was only a few blocks away at North Hollywood Jewelers. Everyone in the store with me was frozen in a state of shock. It sounded like someone had started a war.

No one within sight was safe. Police and civilians alike were shot and hit with armor-piercing bullets, even after taking cover behind cement walls and cars. The bullets sliced through them like hot knives. News crews from the local stations arrived quickly on the scene, as did the helicopters overhead, and the images were broadcast live for over an hour on nationwide television. The two men, dressed in black with black hoods, stood their ground, firing their machine guns at anything that moved. In one of the more dramatic moments of the telecast, a news helicopter reported that they were being fired upon as they pulled away from the range of the high-powered rifles.

Two bank robbers outgunned the LAPD. The police cruisers carried the standard-issue Beretta 92FS 9mm pistols, .38-Special revolvers, and shotguns. The weapons at the disposal of the robbers had much greater range and power—AKM assault rifles, an HK G3A4 battle rifle and grenade launcher, and an M16A1 assault rifle—and their weapons had been illegally modified to make them even more powerful. In the initial moments of the siege, a police officer and a civilian were hit and trapped by the hail of gunfire.

Across the street from the Bank of America sits a Hughes Family Market, a small grocery store, where a couple of the civilians who had been hit were trapped along with about a dozen others. A police officer was hit in the leg and trapped behind a medium-sized tree that acted as cover. He ran the risk of bleeding to death while he was pinned down by automatic gunfire. As much as the officers wanted to get in and recover the injured, they couldn't. Every time an attempt was made, Phillips and Matasareanu pointed their machine guns in the direction of the rescue attempt and fired hundreds of rounds to chase away the would-be rescuers. News cameras captured images of police cars being reduced to piles of junk, riddled with bullets that sent pieces of glass and twisted metal flying, and both cops and civilians scrambled to find cover. At one point during the standoff, LAPD officers went into a private gun store in the area to commandeer weapons more powerful than their standard issue, in hopes of arming themselves with something that could penetrate the body armor the suspects were wearing. The owner of the shop offered anything in his inventory, but they still were no match for Phillips and Matasareanu's arsenal.

As the siege raged on, over three hundred police officers became involved. Throughout the hour-long ordeal, a total of fifteen people were wounded, including ten policemen, as the two assailants fired approximately 1,100 rounds. America watched as dozens of squad cars, fire engines, ambulances, and news trucks filled the area surrounding the bank, held back by just two gunmen. News commentators marveled at how the two could hold off so many and wondered how much longer their ammunition would hold out. SWAT teams started to move in and heavy-assault vehicles rolled into position, when one of the gunmen went to a white car and opened the trunk, reached in, and grabbed more ammunition as if he were taking out a six-pack. It wasn't playing out the way it would in the movies. Hollywood would have these guys running behind cars or walls for protection from police fire, maybe with a couple of cries of "Cover me!" thrown in. The real-life scenario was that these two were standing in plain sight, rarely seeking cover, and in complete control.

Arrogantly, they placed themselves in front of the white single-story bank. Los Angeles had not seen this kind of firefight since 1974 when the Symbionese Liberation Army opened fire on the

LAPD three months after they had kidnapped newspaper heiress Patty Hearst. That event was also broadcast live across the nation, but the assailants were hiding out in a house in a quiet neighborhood, not standing in broad daylight for the news cameras to capture and the world to see.

The gunmen stood their ground and looked almost casual as they fired at anything that moved. Fourteen people were still inside the bank during the shootout and found sanctuary in the bank vault. After about forty-five minutes, the gunmen appeared to be making their getaway. Phillips got behind the wheel of the small white car that held their weapons and ammunition in the trunk and slowly drove out of the bank parking lot, while Matasareanu walked next to the car. Both fired their weapons randomly as they attempted their escape. After about a block and a half, Phillips pulled away and left his partner in crime to fend for himself.

Matasareanu continued to fire his machine gun at the police until it finally jammed, then he took a Beretta 92F pistol he was carrying and fired it at police. This was a break for the officers at the scene. They were able to move in closer to get a better shot at the masked gunman because the pistol Matasareanu was firing did not have the range or power of the assault rifle that had jammed. He was shot in his right hand. Even that didn't stop him. He picked up the pistol with his left hand and continued firing as he took cover behind a truck. He made a final move. Initial reports stated that the police killed him, but after the videos were reviewed, it became apparent that the head wound that killed Matasareanu was self-inflicted. It may have been an involuntary response after an LAPD bullet severed his spinal cord as he tried to reload his pistol.

LAPD now only had one gunman to contend with and shot out the tires of the getaway car after Phillips had gone about a block. Phillips got out of the car and tried to carjack a pickup truck. But after he transferred all the weapons and ammunition to the truck he had just stolen, he couldn't get it started, so he fired off his remaining bullets before finally surrendering. The second bank robber, and the only one alive at this point, had multiple gunshot wounds to his legs and was bleeding badly. It was later learned that both criminals had performance enhancing drugs in their systems, which explained why they were able to continue the gunfight even though they had suffered gunshot wounds to their lower legs.

The scene remained chaotic after Phillips hit the ground, and the LAPD was still under the assumption that there were more involved in the robbery, and they combed the area looking for suspects. Reports were coming in that witnesses saw other people involved in the robbery, and officer John Futrell was ordered to stay with Phillips. John was kneeling over Phillips, who was lying on the ground, with his hand on his back, when a photographer snapped a shot that would be seen all over the world with the caption, "LAPD Officer taking into custody bank robbery suspect. The suspect later died." As a photographer myself, I have always dreamed of winning a major award for one of my photos. But it is a matter of luck being at the right place at the right time. The picture of John over the bad guy won the Pulitzer Prize.

As the rest of the LAPD involved in the shootout were searching the area for what they thought were more suspects, John was ordered to stay with the fallen criminal, and he did as he was ordered. He performed first aid as best as he could while waiting for medical personnel to attend to the wounded Phillips. The area was still considered a hot zone, as police were under the impression that more suspects were still in the area; but it turned out that the sightings of supposed suspects were really sightings of undercover cops in body armor. The delay caused by the misunderstanding resulted in Phillips bleeding to death before medical personnel were cleared to retrieve him.

The shootout lasted for almost an hour and caused the closure of nine elementary schools, residents being ordered to stay in their homes, and traffic jams on Southern California freeways for hours as police combed neighborhoods searching for anyone else who might have been involved in the robbery. When the dust settled and the search was over, and reality sunk in—only the two dead gunmen were involved. There were no other accomplices.

The media went into the predictable frenzy covering the story, interviewing every witness they could find, and the gun-control debate got a boost and was back on the news shows for a while. For the LAPD, it was a double-edged sword. The men and women of the department conducted themselves as heroes and did an exemplary job of protecting the public during the shootout by putting themselves at risk. But, just two gunmen also held them at bay. It did not sit well with the higher-ups in the police department that over three hundred police officers were out-gunned by only two bad guys. It did very little to lessen the embarrassment to say that the two were very well armed, and it also did not improve public confidence in the department.

Of course, there were changes in the department's policies on how to handle this type of situation in the future, and the management of the department had to endure public criticism.

But, when Phillips's family sued the LAPD claiming that the police officers' actions or inaction in getting Phillips medical attention on the scene had caused his death, the LAPD was put on the defensive like never before. Now it didn't only have to explain why the shootout lasted so long, but it also had to defend the actions that prevented Phillips from receiving the medical attention he needed and that may have saved his life.

John Futrell was a veteran of the force, well-respected and liked by his peers. He did volunteer work at local schools, speaking to kids and showing them baseball cards adorned with his photo. He was committed to the LAPD and the public. But LAPD needed a scapegoat. As I said before, John was following orders when he stayed with Phillips as he lay on the street bleeding to death. He had no way of calling for medical assistance for Phillips while other officers combed the area because of false tips that other criminals were involved and still in the area. If he had done anything other than what he did, he would have violated direct orders from his superiors. But he was being a good cop and respecting the chain of command, while putting his life at risk to protect the public.

Shamelessly, the department blamed him for Phillips's death. They fired him for following orders, claiming that he did not follow procedure to help the wounded bank robber. The case that was brought against the department was also brought against John personally, and the LAPD hung him out to dry. Because of some skillful legal representation by his attorney, Terry Goldberg, the department paid out a fifty-thousand-dollar settlement on John's behalf, but his career was still over.

John Futrell did not give up without a fight. He took the department to court for wrongful termination, but the case was too big and the consequences too high for him to have a fair shot at an honest hearing. Records were conveniently lost and witness testimonies were far different than they were at the time of the shooting, and his fight to have justice from city hall failed. It wasn't enough to say that the man who had tried to kill hundreds of innocent people died because of a situation he had created by his criminal acts, or that the crime scene was in so much turmoil that it was impossible to get medical help to Phillips in time. No one wanted to take into account that there were still plenty of reasons for the LAPD to believe it was unsafe for emergency medical personnel to enter the crime scene. Over a thousand shots had been fired, fifteen people were wounded, and no one had any way of knowing for sure that there were not more criminals with high-powered weapons in the area, waiting to take the next shot at anything or anyone that moved. The death of Phillips was unfortunate. But laying the blame on an officer whose sole purpose for being there was to protect the public is the real tragedy.

I became affiliated with John at a time when the LAPD was starting to take the lawsuit more seriously. Getting his story to news outlets and landing him a movie and book deal was my intent. The shootout itself was enough to drive the story, but what was happening to John is what would have made it interesting *and* important. We hit a wall. No matter what I tried, I couldn't drum up enough interest to get the project off the ground. I couldn't help but wonder if the shootout had taken place in Atlanta, Reno, or any other large city that wasn't near Hollywood, if we would have had better a chance.

It is pure speculation on my part, but the Hollywood machine has to deal with the LAPD on a daily basis, and that may have proven to be a deterrent for fear of future reprisal. As far as getting the press that could make public the injustice being laid upon him, it also fell on deaf ears. When it came to the story, the reporters and news directors found it easier to put the spin on the lawsuit filed by Phillips's family, than on a lone police officer who was doing his job and now trying to stand up for his own rights. John Futrell's only crime was that he was the one cop ordered to stand over Phillips. He was doing his job, and doing it well. The thanks he received from the department to which he had dedicated much of his adult life was to have the blame for the bank robber's death pinned on him and then being fired.

The shootout itself did make it to the screen, via public domain. A movie called *44 Minutes* was made, dramatically chronicling the shootout, and the opening segment of the blockbuster *SWAT* was a carbon copy of the fateful morning the robbery took place. The History Channel also did a piece on the shootout, as did a couple of other production companies, but not one of them ever mentioned John Futrell being the scapegoat for the LAPD. John's attorney, Terry Goldberg, tells me that John is still a salt-of-the-earth kind of guy. He is not bitter and still believes in the principles that made him a great cop. Those qualities only make it harder to deal with the fact that his story never got the publicity it deserved.

There were unfortunate similarities between John's situation and a news piece I was involved with in 1993. The story did have a happy ending, just not as happy as I would have liked.

In October 1993, I was involved as a producer in the development of a feature film project called *Fly Girls*. I partnered on the project with Wendy Finerman, a successful movie producer. Wendy has *Forrest Gump* and *The Fan,* among others, on her list of successes. We were working together on a story that was based on a book called *Women Pilots of World War II* by Jean Hascole Cole, and we wanted the screenplay to be as accurate as possible. I worked with the military to procure the book rights and find the right woman pilot to star in the film. A military liaison I worked with gave my name and number to Lorie Durant after he found out about my involvement as a journalist and producer in TV News. Lorie called me the next day to tell me that her husband was being held captive in Somalia.

Somalia is one of those small East African countries that most Americans would never have heard of, much less taken an interest in, if it were not for the fact that we had a military presence there. Like so many small countries with names most have never heard, the country had been in a state of military control for years, leading to civil war. For Somalia, it began in 1977. Power struggles between the military and the warlords who wanted to take control resulted in bloodshed that became a fact of life. The United Nations launched a humanitarian effort to alleviate famine conditions in 1993, and the U.S. military was also sent for support. It was not a war, officially; U.S. forces were there as peacekeepers.

On October 3, 1993, an assault team of Delta Force, America's most elite antiterrorist military unit, was conducting a mission in Somalia, backed up by nearly 100 U.S. Army rangers. Helicopters were to fly deep into the city of Mogadishu and capture two top lieutenants of a Somali warlord. The objective of the mission was to weaken the infrastructure of the powers behind the bloodshed of the civil war. For the elite fighting men of these highly prestigious fighting units, it was a fairly simple task—that was until surface-to-air missiles shot down two Army Black Hawk helicopters in the depths of the city. What was to be a fairly routine mission turned into a struggle for survival for the flight crews of the two downed aircrafts and an embarrassment for the leaders in Washington D.C. The civilian population in Mogadishu was heavily armed and put up one hell of a fight for the rangers who were trying to retrieve the downed flight crews. Casualties were adding up on both sides.

Lorie Durant's husband was one of the pilots shot down, and she pleaded with me to use my influence to pressure the press to cover her husband's situation. It was her hope that the media attention might help motivate the government to secure her husband's release. At first I felt that this situation might be a little out of my league, but I gave it a shot. After hearing the sense of urgency in Lorie's voice, I did not have a choice.

The calls I made to the major networks produced no results at first. The situation was already receiving attention from the media, and the programmers at the major networks felt that the story of just one helicopter pilot would not justify the airtime needed to tell his story. It was discouraging, so I decided to call on some of my contacts at the news magazines. I hit the jackpot. It took a lot of strategic media phone calls and hard-pressure selling, but I was able to show how the power of Durant's story could enhance the picture of U.S. involvement in Somalia—enough so that a photograph of Michael's badly beaten face was on the cover of *Time* magazine with the caption, "What in the World Are We Doing in Somalia?" What was once a little-known country in some far-off corner of the world became a large political issue, and the Clinton administration suddenly found it very important to retrieve the American pilots and flight crews that were being held captive there.

The release of the American airmen received some attention from the major networks; but for the most part it was a story for a few days, then the media monster was on the hunt for the next big thing. The political unrest in Somalia and the events that accompanied it never really captured the public's attention and was of little use in the networks ratings wars. I knew that Michael's story had the potential to become a great movie and book.

After Michael's rescue and return, I traveled to his home in Clarksville, Tennessee, a small town that borders the massive Army installation of Fort Campbell, where Michael was stationed, along the Tennessee-Kentucky border. I was producing a news magazine segment on Michael, and I did not know what to expect. Though he had just been through a horrifying experience, he still wanted to tell his story as much as I wanted to hear it. As I spent time with them in their home, I found the Durants to be very lovely people. I listened to Michael recount his ordeal, and was riveted by his nightmarish memories.

Details about day-to-day life while on assignment, up to and including the moments when his helicopter was shot down, were stunning. I knew I was right in thinking that his story had potential in the entertainment aspect of what I do. First, I acquired his news rights and got his story on some of the bigger network shows. The attention was everything we wanted to position ourselves to deal with Hollywood and the publishing houses. In the midst of the publicity, Michael endured disappointment when he learned that the White House had overlooked him for any medals in recognition of his courage under fire. Two of the rangers who were killed trying to rescue him were awarded the Medal of Honor posthumously. He was hoping to be considered for a Medal of Honor; he was not even mentioned.

A violent combination of the crash and his treatment as a POW had broken his back, and it was doubtful that he would ever fly again. His future in the military was uncertain, as was how he would provide for his family. He risked his life for his country and his reward seemed to be only uncertainty. He gave me his dog tag and POW bracelet, which now hang in my war room in a shadow box with the cover of *Time* magazine. Military dog tags come in pairs. One is on a chain that hangs around the neck, and the other is on a short chain that hangs on the long chain. The purpose of the short chain is to put on the toe of a fallen solider for easier identification. As I look up at Michael's dog tag, I realized how close he had come to having the short chain used.

Michael was an unsung hero that Washington had quickly forgotten. The injustice drove me to get a deal so at least he would have some financial resources to help his family while telling a story that needed to be told. And we came close. We were within weeks of closing a deal with Showtime to do a film on Michael's ordeal, but then for some unknown reason the deal fell apart.

Reports were surfacing on worldwide news agencies questioning U.S. and United Nations involvement in Somalia. If the UN was there for humanitarian relief from famine, and the military was officially there for support of the UN, why were we sending in armed troops and conducting combat missions? The reports were also bringing to light the death tolls of the battle. Nineteen U.S. soldiers were killed in the fighting, and thousands of the Somalis were killed in the cross fire between U.S. troops and the warlords. In some international circles, it was argued that the action taken by the Somalis in shooting down the Black Hawks might have been justified. In the days before the downing of the U.S. helicopters, two UN attacks had killed several men, women, and children. It appeared to some that they were just defending themselves.

What I had hoped to bring to the public was a story of how a real hero survived and to illustrate just how wonderful the human spirit can be when enduring great hardship—not a story with a political backdrop that was becoming unpopular. It would have been a moving film.

Eventually a movie was made about the incident. A film based on the book *Black Hawk Down: A Story of Modern War* hit theaters in 2001. Although Michael Durant was portrayed in the film, and he was a consultant, it centered more on the military operation than on any of the individuals. The producers of the film were smart; they got the cooperation of the military in its production. It was also this involvement that brought the film *Black Hawk Down* one of its biggest criticisms. The terms by which the military agreed to participate in the film production included the right to exclude anything they felt was not appropriate. In return, the military offered aircraft, training for the actors on army bases, and consultants for the writers and directors. It is close to impossible to tell a story in an unbiased manner when the U.S. military and its policies are a central theme and the U.S. military and its policy makers are in control of the film's content. Nevertheless, the film went on to be a blockbuster and won two Oscars in 2002 for Film Editing and Sound Editing.

Michael Durant will always be a hero and worthy of all the medals the military can throw at him. I just wish I could have found a way to have his story, not the military's, shown. It would have been unforgettable.

Doctors who are willing to inject themselves with the AIDS virus to prove they have a working vaccine, a police officer that became the scapegoat for the nation's second-largest police department, and a lone army helicopter pilot who was shot down over a small African country in 1993—these are not the types of stories that people ask me about when we first meet. But they should, because they are important stories and they will never learn the truth about them in the normal news and entertainment channels. Like I said before, I have enjoyed many accomplishments in this crazy business, and I have a lot of good memories and relationships that will be with me for the rest of my life. But the stories that have touched me the most involve people who are inspirational and deserve to be called heroes—oftentimes, these are the stories that never get told. These are the casualties of the ratings war.

11

Too Close to Home

Say the word *newsman,* and most people conjure up images of someone hard as nails, lacking any compassion, hell-bent on getting the story with no regard for the people involved, or other traits that make news people different than the rest of us. Although I admit that I have met many people who fit this description, most of the people I have worked with in the industry are kind and compassionate. Objectivity is essential. The relationships I develop with my clients are not the same as those they have with the reporters. In many cases, I spend real time with my clients, and relationships form that often turn into friendships. To be truly objective, I have to be able to look at a situation from the outside, the way the news will report on it. It is not always easy.

If I were asked what one trait makes me so productive at what I do, I would say it is that I am a spiritual person. If I were asked what one trait I wish I could change, sometimes I would answer that it is being a spiritual person. I don't mean to be coy; it really is the truth. I am good at what I do because I can quickly see what stories will touch the American public emotionally. The only difference in a story that receives a quick mention at the end of a newscast and one that will be a ratings sweep story is how deeply it touches the people who watch it. The more emotion a news piece brings to the small screen, the more people will watch. It is my spirituality that helps me see the stories that will touch viewers. It is also that trait that can make this business draining.

I love people and truly care about their feelings. Because of the work I do, I see people at their best and at their worst. I have lost count of the times I have sat down and cried with my clients/friends as we experience together the fear their story represents and what they have endured. News loves tragedy, and many of the stories I become involved with are just plain sad. People who have lost loved ones in violent and senseless acts, a husband and wife who had to bury a son because of a freak accident, a family who lost everything to a natural disaster—the examples are endless. When people come to me, they often see me as more than just a front man to the media monster. I often become a counselor or the one who will tell them everything is going to be OK. What you see on the TV screen I designed to capture your attention quickly. I also deal with what you don't see: the pain and devastation that many times accompany the stories you want to watch. Knowing this before made me think I would be more prepared to handle the unfair blows that happened in my own life. I was wrong.

It was August 12, 2004, my birthday. My son and I had just gotten up and were enjoying our morning together, when the phone rang. When the phone rings in my house it is a little different than most other homes. For every call that is a friend or family member just calling to chat, there are at least twenty-five calls that are work-related. My work brings me some very interesting calls. I never know if the person who is calling has the next huge "Oh, my God" story or not, so every time the phone rings I tense up a bit, ready to go to work at a moment's notice.

This time the call was from a police officer that wanted to speak with my son, Sean. Sean is a great individual, so I knew he wasn't in trouble with the law. My mind was spinning, "Why in the world would the police want to speak to Sean?" I was more bewildered than worried; it didn't make sense. After a couple of moments, I explained to the officer that my son was a minor and that if he needed to talk to him he would have to let me know why first. The officer told me that Shannon, my ex-wife and the mother of my three kids, had been in an accident. For a moment it felt like someone had wired me up and run an electric shock through my body. Shannon and I had been divorced for years, but still had a co-parenting relationship. I still cared for her as a wonderful person. More importantly, she was a good mother to my children, and the thought of her being harmed in an accident was frightening. I was in shock when I half-stated, half-asked, "An accident . . ." Sean came running to me when he had heard my tone and the word. "What's wrong, Dad?" he asked with concern in his voice.

The officer explained that Shannon had been at work, when a stalker had shown up unexpectedly. I could tell that the officer was going to tell me something horrible because of the timbre of his voice. It is never easy giving bad news to anyone, let alone a stranger. "Mr. Garrison, your ex-wife has been shot. Four times to be exact, and she is being airlifted to the hospital right now." Now it felt as if someone had turned up the amperage in the cord that was running through my body. I asked what her condition was, if she was OK, could she speak, where was she hit by the bullets? Somehow, I knew that my ex-wife wasn't dead. I just felt that she was alive. I also felt the pain and fear overwhelm me as I realized that my kids were going to have to live through something I might have been covering, had circumstances been different. I have had to do this in other people's lives. If someone had asked me a moment before the phone call if my work helps prepare me for something like this, I would have said yes. Now I know the truth; it makes it worse. I have seen so many horrible situations like this before, and I know by working with people who have lived it that there was nothing that I could do to spare my children the pain and fear that would follow.

I was so proud of my kids; they really held it together. Sean was shaken but strong as we started to contact his two sisters. I called my oldest, Jaime, and tried to break the news to her as gently as I could. But there is no gentle way to tell your daughter that her mother has been shot. I quickly explained all I knew about the shooting from what the police officer had told me, and then she was on her way to my house. My son contacted my younger daughter, Lindsay, who then was on her way to the hospital to see her mom. To this day, I really wish she had waited to go with me. She is a bright and very capable young woman, but I was still very worried about her and wanted to be able to hold and protect her.

Shannon was airlifted to an area hospital that had an emergency room better equipped to handle shooting victims. The kids and I rushed to the hospital, praying for the best. Although the drive seems like a blur to me now, I do remember realizing that whoever had done this horrible thing was still on the loose. I remember thinking, *Is he going to come after my kids?* It had been hard enough hearing the news that my ex-wife had been shot and not knowing what her condition was. I also wanted to do everything I could to protect my children from harm. I knew that Sean and Jaime already had way too much to deal with, so I kept my fears to myself. Silently, I started to formulate a plan to ensure that my children would not be harmed.

When we arrived at the hospital, it became even more frustrating. It is the hospital's policy not to list the names of gunshot victims. We had no way of knowing if we were at the right hospital or if the police officer on the phone had given us the wrong information. I even remember a fleeting moment of relief. *Maybe it was all a mistake! Shannon isn't shot; she is sitting behind her desk right now working.* The relief was short-lived, however. A passing nurse saw the fear in my children's eyes and was kind enough to take us to their mom.

This cannot be happening, is all I could think when I saw Shannon in the emergency room. Our children were standing beside me as we looked at her, and nothing could have prepared us for the impact it had. When people get shot in the movies or on TV, the portrayal is very rarely even close to what really happens. There is not an actor in the world that could convey the fear or pain in our eyes. Still, she was alive. She had been shot in the chest, head, neck, and arm. It really was a miracle that she didn't die at the scene of the crime; but she was a fighter. The Irish-Italian temperament that had made her the courageous person she was was saving her now.

I left the room so the kids could be with their mom. I spoke to the police about my fears that the person responsible for the shooting might want to bring harm to my children, and their response was less than comforting. "Well, Mr. Garrison, you really should do what you can to secure your home, you never know with cases like this." I made a mental note to make sure the house was locked up tight, but I knew I wouldn't be getting much sleep until the stalker had been caught. I also wanted to protect my kids from the press. I was not about to let them become victims of spin. I acted as a barrier to make sure that no one from the media would have access to any of them or their mother. The reports still ran that night and, except for the details of the shooting itself, were wrong. I shook my head when they got her nationality wrong and got just plain mad when they inaccurately reported that it was because of a family dispute. Sometimes, I really am ashamed of the industry I work in.

At the forefront of my mind was that the man who shot Shannon was still out there. At first I thought it best to let the police do their job and find him, so I could do my job and be there for my kids. But the next evening, my neighbors spotted a man fitting the description of the stalker in a tree next to my home. I decided to take action and use whatever I had to get the guy caught. I have been in the news business for a long time, and even though I did not want my kids to face reporters, I had no problem using the industry to get the man who shot my ex-wife. I e-mailed my friend and colleague Diane Sawyer, and explained what had happened. I asked if she could use her clout to get the suspect's likeness from police sketches on the air in my local market. Her response was quick and gracious, and she offered whatever help I needed. It was nice to know that my peers stood by me.

The police called and informed me that the stalker had been found in his car, dead from a self-inflicted gunshot wound. Part of me was enraged. I wanted my family to see justice, and this guy had taken the coward's way out. But I quickly thought better of it, knowing that the nightmare would end more quickly this way. I didn't have to fear for my children's safety anymore and it would give us a chance to heal from the tragic event. Ironically, my ex-wife was shot on my birthday, and the stalker took his life on my daughter's birthday. I guess we will never forget those events or celebrate our birthdays in the same manner again.

Shannon gave it her best, but passed away in the months that followed the shooting. I have seen firsthand how violence of this type reaches into people and scars them, but I have also seen how a family can be pulled together because of it. My kids and I have grown closer and our family is tighter than ever. They miss their mother terribly, but are learning to cope with her loss.

I have always been sensitive to my clients' feelings, but I must admit that after the experience my family endured, I am even more careful now. Before, I could only imagine what it was like to be in a position like that. I have lost loved ones before, some suddenly, but nothing like what happened to Shannon. Now I do not have to imagine; I know. I have lived it through my children's and my ex-wife's pain. I'm not sure if it makes me a better newsman or not, but I can tell you this: There is absolutely nothing I wouldn't give for my kids to have their mother back.

12

The Road Ahead

There is much to consider when contemplating what the future will hold for this complex and essential industry. The news media has an awesome responsibility to keep the public informed about important issues that have a direct impact on their lives. Things happen faster than ever, and studies show that people have less time to read newspapers and magazines because of their busy schedules. So, TV news has become the way most find out what is happening. In fact, I believe that the TV news media's responsibility to keep the public informed is the backbone of the democratic process. The news media is big business, usually controlled by the ratings wars and financial influences that all businesses have to contend with.

Seeking to grasp what the future may hold, I have to look back at my career and see how things have changed over the last twenty-five years. When I was a producer at the Dick Clark Film Group, I learned about taking a news story and developing it into an entertainment product. I earned my journalistic credentials while with Time Warner and then ABC. I was supplying ABC with stories and was told that there was a problem between Diane Sawyer and Barbara Walters as to who would get my next story. The network decided to merge its two popular news magazine shows into one. *Primetime Live* joined forces with *20/20*, putting them both on the same show. Dick Wald was one of the heavyweights in the news department at ABC and flew me out to New York for me to join their team. The concern that having two of the most powerful news celebrities on one show could cause a potential internal conflict was being realized, so I was brought on board to make sure there were no shortages of news pieces for them both to cover. It was at that meeting that I knew I had made it as a newsman. As we wrapped up our meeting and I was walking toward the elevator on the second floor, I looked down through a glass wall at the news studio below and saw a sign that read *World News Tonight*. Peter Jennings was seated at his anchor chair, and he looked up and gave me a smile. It is a moment that I will never forget.

While with ABC, I worked with the senior news producers and got to see firsthand how the process of deciding what was going to be aired worked. At that point in the news cycle, the main force was the news. The ratings showed that the public wanted to know more about the important news stories and didn't really focus on the tabloid type of programming that would later capture its attention. Political stories and detailed reports on the stories behind the stories were the main subject of the news shows of the time, and pure raw journalism was the rule.

Another of the shows I worked extensively with at Time Warner was *Extra*. The syndicated show leaned toward the types of stories that were easier to spin for dramatic effect. Even though the subjects they covered tended to be more like tabloid stories, their main focus was still the hard news. The smaller syndicated shows operated basically the same as their big network counterparts. A news producer or booker would find a story, do all the research, and verify all of the facts. Once all the information was gathered, they outlined a script of how the news segment was to air. It included what video footage would be used and in what order the story would be told. If there was going to be an on-camera interview, the producer outlined the questions that the on-air journalist would ask. Most people think that when they watch an interview on a news show that it is spontaneous. Only if it is live. For the most part, the interview process is calculated and filmed in a way that makes it easier for the producer to edit the final product before it is aired.

No matter what the format, or if a show was a network heavyweight or a smaller syndicated show, the focus was on delivering the news and different aspects of the story. Ratings went up if a news program could divulge unknown information about a headline story, and that is where the competition between the news agencies heated up. As time passed and more news sources became available for viewers to watch, the competition became fierce. News programmers had to become more creative in the way they presented the news, and some of the new networks started to use spin and report in a more slanted way. This new type of reporting captured the ratings. The rest of the industry noticed and started to change the way they presented the news. Reporting the news drifted away from telling purely what happened to offering implications and opinions of the events. When a new format was introduced that dedicated entire shows to one story, it changed the way the news was presented even further.

Trends on the shows that give in-depth coverage of major

news stories have brought a kind of shock TV to the air. Hour-long blocks of airtime are devoted to analyze every little particle of a story. This trend has created the need to have "expert" consultants go into detail on the air about their particular subject. I have experienced that with the Natalee Holloway case. I mentioned previously that I had acquired the news rights to two of the jurors from the Michael Jackson case, one of whom said that she and a fellow juror were "pressured to find the pop star not guilty." So-called experts now debate her statement on air. Years ago, the only people the news agencies wanted to put on the air were people directly related to the story. Now the competitive atmosphere of the industry opens up the range of who will be in an on-air interview. I received countless calls from a variety of news outlets to go on-camera and be interviewed about my client's part of the story. Years ago that would have been unheard of, but with so many shows grasping for any way to capture viewers' attention and the ratings, more and more people are being interviewed.

Another trend is having experts in a subject go on-air and help us form conclusions about the story. Experts about the law, economy, fashion—you name it. The industry's gradual shift from giving us the headlines to going deeper and deeper into a news piece has created a need to have people who make a living by telling us what to expect or how we should review the facts. In the old days there was an unspoken professional courtesy that ruled the conduct of commentators or consultants. Today, if a shouting match erupts because of a difference of opinion, well, all the better for the ratings.

Over the last few years, the type of news the public

wants has also changed. Shows that dedicated their efforts to mainstream news were losing market share in the ratings war to broadcasts that emphasized celebrity news and gossip. Years ago, Barbara Walters started a very successful series of prime-time specials that featured interviews with celebrities. She still does it now, sometimes, when it is warranted and on Oscar night. Her show, *The Barbra Walters Special,* gave a glimpse into the behind-the-scenes lives of Hollywood royalty.

Today, the public is not satisfied with a report every few months about a couple of different stars. CNN is just one example of how the impact of the public's desire to know more about the lives of the rich and famous has changed the news industry. The Headline News channel was once a true source that anyone could tune into at any time of day to catch up on daily events. But the daily events were not enough to keep the cable channel competitive for the market share, so they now dedicate their prime time to shows that focus on entertainment-industry news.

Attention shifted from hard news stories to the subject of the entertainment industry. Once again, the news industry is big business, and when the ratings show that more people will tune in when reports of celebrity gossip fills a broadcast rather than hard news, the programmers are going to go after the audience. And the audience, according to the Nielsen Ratings Reports, is more interested in what movie star marriage is falling apart than what is happening in Washington D.C. The syndicated show I mentioned before, *Extra,* had to change its format to cover exclusively the news and gossip on the entertainment industry to stay competitive in the ratings war. In fact, a couple of the show's counterparts are no longer on the air; they did not shift their reporting focus in time.

Change is the one constant in the news industry. Whether because the general public is more educated or just has a shorter attention span than it did years ago, the programmers in the news media have had to realign their thinking on how to present the news and also on what news to present. In the past, news anchors read the headlines into the camera without much pizzazz. Anchors today are much more attractive and put a lot more personality into the delivery of the headlines. Stories that would not have been covered comprehensively years ago are today's top news pieces. If the Scott Peterson case had taken place ten years before it did, I really do not think it would have received the attention the media monster bestowed on it. There is a lot more competition with the influx of more news sources, and the need to fill airtime has created the need to report on stories that would not have received much attention in the '80s.

When disaster strikes or something really significant happens, the news shows will always focus their attention on the events of the story. Hurricane Katrina filled the news airways for weeks, replacing only temporarily the in-depth coverage of the big story of the time, Natalee Holloway's disappearance in Aruba. What will capture the news audience one day can change in a matter of moments, as disaster or tragedy on a large scale takes precedence. But when the smoke clears and the major events are milked for all they are worth, some other story about a disappearance in some far-off place or the murder of someone you have never heard of before will be needed to fill in the air time—at least until the next big news event takes place.

News media will always focus its attention on the famous when they are suspected of committing a crime. The O. J. Simpson case taught news programmers that the public will tune in to find out all the details and speculation on a celebrity suspect. Michael Jackson, Martha Stewart, Robert Blake, and any other public figure who is caught up in an investigation or trial is guaranteed to have the situation covered intensely. It is just impossible for the news outlets to resist. The coverage will not be limited to the programs that focus on the entertainment business; it will also be covered in the mainstream news cast. And coverage will always add to the ratings.

One day, I picked up my teenaged son from school and asked him about the future of the news. I asked him what he thought about television news today, and his answer came as a shock. He said it was all spin the way the focus has shifted from what is really important to things that nobody should really care about. He used the example of how there were so many reports about Britney Spears and other gossip filling the news shows. He felt that the industry should be focused on things like the homeless, world hunger, and the suffering of people who are starving in Third World countries. He went on to say that it was hard for him to trust what he saw on television news when they did report on the more important stories.

When I think about the conversation with my son about the news, it brings to mind some other things that will have an impact on the way the news is reported. Children of today are much more sophisticated with the technologies that fill our world now. My four-year-old grandson, Dylan, blew me away with the skills he learned this past summer in a computer class he completed. I could not keep up with him on his Sponge Bob game. When I consider how the developments of the cameras and editing equipment have evolved, I can see how the way the news agencies of tomorrow will gather the information they bring to you in a different way also.

Before, when reporters went into the field it was a fairly large task. It was a reporter and his crew. The cameraman and soundman accompanied the reporter in large trucks that held the equipment needed to record the news events or interviews. It was a complicated and expensive process. Today, many of the news agencies have switched from the big news vans full of technicians to one-man shows. Now a reporter armed with a digital camera and a laptop computer can do the same job alone that once required a team, making it easier and faster to get the information to the public.

The sight of a stranger holding a huge camera pointed at you from behind a reporter holding a microphone in your face as he fires off question after question is very intimidating when compared to one person asking you questions using a camera the size of a cell phone. I believe that with the new technologies and the need for fewer people to be involved, the interviews we see on the nightly newscasts will be vastly different in the future. I also see networks pooling stories to save money and time. It will be easier for people to be more candid, and the more relaxed atmosphere will make it easier for the reporter to dig out more information from the subject of the interview.

It wouldn't be the first time technology had an impact on the way news is brought to us. Think about the satellite and how it has affected the way we receive our news. One of the best illustrations would be the first war in Iraq, Desert Storm. Unlike before, we did not have to wait to see the images the war produced. The news programs were filled with real-time pictures of exploding bombs and pictures of journalists running from harm's way as sirens blasted in the background. The visuals of war were immediate.

Increasingly new developments in equipment will also make reporting the news look different. In the past, the feeling when watching the news reports was that viewers were on the outside looking in. In the future, it will feel more like viewers are part of the story, almost as if they are participating in it. The job of a reporter has become much more portable, enough so that news people can get right in the middle of the story while it is happening. As I watched the reports of the aftermath of Hurricane Katrina, I was amazed how many of the reporters were actually right there during the rescue efforts. In fact, one report that stands out in my mind showed reporters actually helping to rescue animals from submerged homes in New Orleans. The impact is much greater when watching news stories from a perspective of being right there where one can see the expressions on the faces of victims, instead of through the lens of a helicopter-based camera from five hundred feet above.

With the new portability of the equipment needed to deliver the news and the willingness of many news people to put themselves at risk to bring viewers the scenes, another line will become blurred. As reporters get closer and become more involved with the story as it unfolds, the lines determining what is a story will be crossed, and the news coverage will become part of the story itself. Where before we felt like observers, in the future we will feel like we are participating.

The more intimate feel of the news in the future will cause a greater impact on the major issues that come before us. In the past, it was challenging to help the viewer relate to a situation. In the '90s I procured the news and film rights to a book involving dolphins called *Behind the Dolphin Smile.* Rick O'Barry, the trainer of Flipper, had come to realize that the way aquatic mammals were being treated in captivity and how they were being used in the military was truly inhumane. The U.S. Navy was training dolphins to clear mines. O'Barry also felt it was detrimental to the dolphins to be kept in small holding tanks. He was correcting the problem he created by telling the truth. His passion to force changes on the treatment of the dolphins was contagious, and I decided to help him get attention from the news media to help his cause.

Back then, we were pretty limited to news conferences and press releases to get attention for the situation. If we had tackled the issue in the future with newer equipment, we could have painted a much more vivid picture for the viewing public of what the true situation was. We could have put the viewer right in the pool with the dolphins and pulled them into the story by showing the emotions and behaviors of these loving creatures. Although we were successful in changing their treatment by raising public awareness, it would have been much more effective if we had today's technology available to us. It was a kick going to Sea World and watching Rick picket them because he felt it was damaging to the dolphins to be kept in small chlorinated tanks. I then became part of the news piece, picketing with him.

With the ability to bring news pieces to the air quicker and easier, the question of our privacy will be raised. The First Amendment will always protect the right of the press to bring you the news in just about any way they see fit. But, with it being easier to conceal video- and audio-recording equipment, the difference in reporting the news by investigative reporting and straight-out spying will become an issue in the future. The American public enjoys its right to know, but how much of their privacy they are willing to lose in the process of bringing the information to them has yet to be seen.

As far as how future news programmers will decide on what to report and how to report on it, that is up to you. The changes in the news industry that we see today are the direct result of what news program you decided to get your information from. Many of us love to complain about how the news has become less personable and more tabloid in nature, and how the use of spin is out of control. Not a day goes by that I don't hear someone complain that the news agencies do not spend enough time on important issues and focus too much energy on the fillers and fluff pieces. People claim to be outraged as they talk about a news show they watched the night before on which the interviewer badgered the guest, or how an outrageous point of view angered them. There is criticism of how too much attention is paid to a single murder, celebrity crime, a missing teenager, live car chases, or any of the other subjects from a long list of news stories that fill the airwaves. The very same reason that this type of news is aired today is the same reason that stories will be chosen for the newscasts of tomorrow: It is what the majority of people wanted to watch. Even on Internet dating sites, people list that they are news junkies. That is what makes our business thrive.

Though the Nielsen Ratings System takes its information from only a very small amount of the public, it is the standard that all news programmers have to live by. Ultimately, the main job for the people who control what you watch on television news is not to inform you, but to make sure that you watch in the first place. If it takes bringing to the air the sensationalized stories that have little impact on what happens in our day-to-day lives, then that is what they are going to air. If having more attractive people deliver your news in a way that helps you form an opinion of the story is going to get you to tune into a news program, then that is what they are going to do. The news industry is driven by the ratings, and viewers are what drive the ratings. Basically, viewers will control the future of the news media. But there may be another way to voice ratings votes in the future.

In late 2005, it was announced that CBS and other major networks, that up to this point relied purely on revenue from advertising dollars, are going to make available their more popular shows on a pay-per-view basis. For years, the networks that broadcast their signals free of charge have been envious of the income that their cable counterparts were able to create. It is marketed as a convenience for people who cannot catch their favorite TV show while it is being aired. For as little as 99 cents, they can have the show on their living room TV anytime they want, as long as it is after the initial broadcast. This will create new possibilities for network programmers, both news and entertainment, to gather information from the pay-per-view data regarding viewership. I have a lot of friends in the industry that are in a position to decide what is going to be aired with regard to television news, and more than a few are a little skeptical of the Nielsen Ratings System. I think the day will come when pay-per-view programming will involve the news media also. The figures will either provide new insight into what we watch or confirm that the ratings system currently in use really is accurate.

The short of it is that news programmers will air what is in demand. The more you watch a particular show, the more valuable the advertising space will be, and they will continue to broadcast what has proven to be financially successful. It's just business, plain and simple.

Watching reassuring faces like Walter Cronkite and Peter Jennings delivering the news will be a thing of the past. Somewhere along the way, viewer loyalty fell away from television news. In the old days, the three major networks all reported pretty much the same information. The only real way to compete for loyal viewership was for ABC, CBS, and NBC to have the most engaging anchor deliver the news to you. They knew that in the market of the time it wasn't so much how or what news was delivered, it was who delivered it that got you to tune in. Times have changed. There are still a few examples: Diane Sawyer, who I am fortunate to call a friend and for whom I have a great deal of respect. Oprah is a spiritual woman and a class act. Barbara Walters is a legend in her own right. The news business has become more insensitive. Unless an anchor is willing to step outside the box and spin the news to capture your attention, thus risking being inaccurate, there is little reason for you to keep your television tuned in to his or her show.

The day may come where the public becomes fed up with the spin and the tabloid type of stories that take up airtime today. Maybe my son's generation will demand more from their news agencies and can expect that the focus will be on the truly important issues of their time, instead of the celebrity gossip that currently rules the airways. I believe in cycles, and eventually we may get back to basics with class reporting and less sensationalism.

As far as my involvement, I feel secure that the public will always have a hunger to know more about the "Oh, my God" type of story I have enjoyed bringing to the news media, and to you. I am a rebel in the news business. No one tells me what to say. I am not owned by anyone, and therefore I will always tell the truth and be truthful to my clients. No matter what trend the future of news programming brings, there will always be those stories that capture your imagination or just make you want to shake your head in disbelief. Some of the stories will be important, some won't.

I will always be on the lookout for the stories that not only capture your attention, but are noteworthy enough to make a positive change—stories that can make a difference in situations that are just not right, simply by making the public aware. I know the future holds plenty of situations that could be made safer or more just, and I will always be looking for those stories. There will be things discovered that the public really should be made aware of for a variety of reasons. Stories will be reported, and it will be up to me to make sure the rest of the story is told. History has a tendency of repeating itself.

As I look back at my history in the news business, I see many of the stories I have been fortunate to be involved with and the positive impacts they have made on society. I am proud that I was able to help them get the attention they deserved. It makes me very excited about the future of the television news business.

Epilogue

"Hey Dad, can you take me to the store to pick up some film?" I hear my son's voice coming from the other room as I sit back in the chair in my war room. "Sure, I'll be there in a second," I yell back to him.

I'm not quite awake, not quite asleep. I am just kind of there. I look around and see the mementos that twenty-five years of my involvement in the news business have accumulated. I can't help but to think to myself that it is a very interesting room. A helicopter pilot that was shot down in some far away, unheard of country; a presidential right-hand man whose private life became too public; a cop who was fired for following orders; the good son of a murderous con woman. These are just a few of the stories that this room has to tell. As funny as it may sound, I come here to relax and to get away from it all. It's my place to escape.

On the other side of the closed door is the rest of my life. My son is waiting for me to drive him to the store, and soon my daughters will be here for a family dinner. I look forward to the times when all of my kids, my grandchild, and my celebrity son-in-law, Michael, are around me with my lady, Lee, who keeps me balanced on news matters and health issues, her son Nolan, and his girlfriend Melissa. We talk about the future and the memories of the past. We also have a chance to help each other heal from the pain of it. If it is like many dinners we have had before, my kids will laugh at me—laugh with me would be a better description. Every once in a while, I talk about retiring, and I always make sure to say *maybe.* They know that maybe means never and that I will be in the news business for a while longer. It is like eating one potato chip; you can never stop. My kids know that this business is under my skin, and if I didn't have it as a part of my life I wouldn't know what to do with myself.

Tonight is going to be different, anyway. I have a new member of the family to introduce to my grandson. A miniature pot-bellied pig is the newest pet at the Garrison residence, and she fits in quite well with the squirrels and the chickens and the turtles and God only knows what else. Her name is Penelope Piglet and, like the chickens, I'm sure she is the only member of her species within miles of my modest home nestled among mansions.

Standing up to stretch from my comfortable easy chair, my son peeks his head in the door. "Ready, Dad?" I smile at his teenage eagerness to finish his latest photo project. "Be right there," I say.

In my mind, I'm thinking it's been a good run. Maybe I should think about retiring one of these days. I can tell my son knows what I am thinking. He knows me pretty well. As I grab the keys off my desk and head for the door to play taxi driver the phone rings, and my son freezes.

I pick up the receiver and recognize the voice of one of my insiders who is always on the lookout for the next big story. My son sits down in the other chair in the room as he sees the signals that I'm getting ready to work. I sit back down in the chair and grab a pen and paper and start to jot down some notes as the insider tells me of his latest find. "You're kidding me?!" I exclaim. As the caller continues to talk I put my hand over the mouthpiece of the phone and silently mouth, "One sec . . ." to my son. He just looks at me and smiles. He knows from experience that this trip to the store has just been delayed. Dad is going to work. But this time, I take my cell phone and he listens in on the new adventure as we go to the car.

I intently listen to the details of the story the caller is giving me, and at the same time feel a little guilty for making my son wait. The story sounds really good, but . . .

I make one more attempt to hang up, but still more information comes spurting out of the phone. We listen to my caller from the White House as we turn to each other and mouth, "*Oh, my God!*"

www.ingramcontent.com/pod-product-compliance
Ingram Content Group UK Ltd.
Pitfield, Milton Keynes, MK11 3LW, UK
UKHW020819120325
456141UK00001B/107